Fundamentals of Electronic Systems Design

Jens Lienig · Hans Bruemmer

Fundamentals of Electronic Systems Design

 Springer

Jens Lienig
Electrical and Computer Engineering
Dresden University of Technology
Dresden, Sachsen
Germany

Hans Bruemmer
Springe, Niedersachsen
Germany

ISBN 978-3-319-85762-6   ISBN 978-3-319-55840-0   (eBook)
DOI 10.1007/978-3-319-55840-0

Printed on acid-free paper

This Springer imprint is published by Springer Nature
The registered company is Springer International Publishing AG
The registered company address is: Gewerbestrasse 11, 6330 Cham, Switzerland

Foreword

It is hard to underestimate the impact that electronic systems have on society. My first portable radio was proudly marked "7 transistors." Today, the electronic system in my pocket has over 2 trillion (2e12) transistors. There are 20 times more transistors in my smartphone than there are stars in the Milky Way galaxy. This unprecedented scale is enabled by the exponential self-fulfilling promise of Moore's law. For over 50 years, the component integration density has been doubling every 18–24 months, while per-device cost went down at the same rate. In all of history, no other industrial law has been as reliable, and no other law has been as influential. Moore's law has fueled the PC revolution in the 1980s, the Internet in the 1990s, the social media in the 2000s, the smartphone, and now the machine learning revolution. New electronic systems extend our senses, helping us see, helping us navigate, and helping us drive safely. The impact reaches beyond the gadgets: Electronic systems affect the way everybody works and lives. It is even fueling political revolutions, for better or for worse....

It is easily overlooked that Moore's law implies that the human designers of electronic systems need to improve their productivity at the same exponential rate. Even a large company like Apple could not afford to double the size of their design team every two years to keep up with design scale. The A10 chip in the iPhone 7 is estimated to have 10× the number of components as the A4 chip in the iPhone 4 just five years earlier. The size of the design team, however, remained roughly the same.

It is no small achievement that electronic systems of this scale can be successfully designed, engineered, and mass-produced. This book addresses the engineering fundamentals behind the design process of effective and reliable electronic systems. Both students and professionals alike will appreciate the contents: The first because it sets up the fundamentals of the entire design process in detail, and the latter because the book brings together state-of-the-art design skills from the extensive experience of the authors.

The first chapters of the book address the architecture and fundamental structure of the design process of electronic systems. That includes the engineering decisions on breaking up the design into more manageable partitions. This must be done in a way that makes assembly straightforward and reliable. It must also be done while taking into account the limitations of tools, physics, regulatory rules, and people.

The main thrust of this book is addressing ways to "tame" the physical effects and control the unwanted side effects of the large-scale integration. The objective is to make the system reliable in production and use, and to make it resilient against external influences. The authors lay down thorough in-depth description of the theory and practice of reliability engineering. After all, it is only as strong as the weakest link.

A significant portion of this book addresses the heat that is dissipated in the electronic system. This is a point where the steady progression of Moore's law poses a true challenge, as the transistor density continues to increase exponentially while the per-transistor power does not decrease at the same rate. To keep the device temperature under control, either the heat needs to be avoided or the heat transfer rate needs to be maximized. The authors present the fundamentals on assessing and optimizing heat flows of electronic systems.

There have been several occasions where products malfunction because of electromagnetic interference. To avoid such design errors, this book provides an excellent description on reducing such unwanted coupling of the system and the environment. The clear set of guidelines and design recommendations is provided to ruggedize the electronic system from the start.

Once an afterthought, minimizing the environmental impact of electronic systems is becoming a major design criterion. There are already billions of electronic systems surrounding us, most of which have a relatively short life span. At the same time, the highly compact and integrated nature of electronic systems makes them harder to open and disassemble. Therefore, even small design improvements matter. An in-depths guide to addressing all environmental aspects during the full design cycle is presented by the authors.

This unique book provides fundamental, complete, and indispensable information regarding the design of electronic systems. This topic has not been addressed as complete and thorough anywhere before. Since the authors are world-renown experts, it is a foundational reference for today's design professionals, as well as for the next generation of engineering students.

Dr. Patrick Groeneveld

Scientist

Synopsys Inc., Mountain View, CA, USA

Preface

> If you have an extreme passion for producing great products,
> it pushes you to be integrated... It takes a lot of hard work to
> make something simple, to truly understand the underlying
> challenges and come up with elegant solutions.
>
> Steve Jobs

We are rarely aware, in our daily use of smartphones, notebooks, etc., that the development of mobile electronic devices started only a few decades ago. After the discovery of the transistor in 1948, the first integrated circuit was built in 1960, followed by the microprocessor in 1971. Then in 1973, Motorola developed the first prototype mobile phone, in 1976, Apple Computer introduced the *Apple I,* and IBM introduced the *IBM PC* in 1981. The popularity in the late 1990s of cell phones and increasingly powerful laptop computers foreshadowed the *iPhones* and *iPads* that became ubiquitous at the turn of the century. We have truly become a society immersed in mobile electronic devices.

The packaging density, i.e., the number of components per unit volume, has increased consistently throughout this period and shows little indication of slowing down. The resulting amount of heat to be dissipated increased as well, putting the spotlight on heat transfer issues. It further became obvious that the reliability, i.e., the function and durability of electronic components, depends greatly on temperature. Another problem identified was the undesirable influence of switching functions, caused by unwanted signals inside and outside packages. These issues came under the heading of *electronic systems design*, which quickly became an important interdisciplinary subdiscipline of electrical engineering.

Since the first appearance of mobile electronic devices, such as the transistor radio in 1954, components have undergone massive development and miniaturization; integrated circuits have reached unheard of complexity levels, and new packaging methods coupled with computer-aided design (CAD) have revolutionized the design of electronic systems. More recently, recycling and environmental requirements were also added to the mix. It is amazing to realize that every smartphone today has more computing power than the on-board computer in

Apollo 11, which transported the first humans to another astronomical object back in 1969.

This book addresses this enormous scientific progress and offers a review of the current state of the art in the development of electronic systems. It is the result of the extensive experience of its two authors in industry, academic research, and teaching in electronic systems design. Its aim is to support the reader with the development and fabrication of modern electronic devices, taking all relevant aspects into consideration with a clear presentation of the underlying technical and scientific principles. The book elucidates a broad range of techniques that have helped keep German engineering at the cutting edge for several decades and will continue to do so for decades to come.

A book of such considerable scope can never be accomplished by one individual. The authors wish to express their warm appreciation and thanks to all who helped produce this publication. We would like to mention in particular Martin Forrestal for his key role in writing the English version of the book. Our warm thanks go to Dr. Mike Alexander who has greatly assisted in the preparation of the English text. We also wish to sincerely thank the following for their support with subsections of the manuscript: Dr. Alfred Kamusella (Sect. 2.6), Dr. Helmut Löbl (Chap. 5), Prof. Stefan Dickmann and Dr. Ralf Jacobs (Chap. 6), Prof. Karl-Heinz Gonschorek (Sect. 6.6), Prof. Günter Röhrs (Chap. 7), Steve Bigalke (Appendices 8.1 and 8.2), and Dr. Frank Reifegerste (Appendices 8.4 and 8.5). Thanks are also due to Nicole Lowary and Charles B. Glaser of Springer for being very supportive and going beyond their call of duty to help out with our requests.

Rapid progress will continue to be made in electronic systems design in the years to come, perhaps by some of the readers of this humble book. The authors are always grateful for any comments or ideas for the future development of the book, and wish you good luck in your careers.

Dresden, Germany   Jens Lienig
Springe, Germany   Hans Bruemmer

Contents

1   **Introduction** 1

2   **Design Process and Its Fundamentals** 5
    2.1   Life Cycle of Electronic Products 5
    2.2   Design and Development Process 6
    2.3   Guidance for Product Planning, Design and Development 8
        2.3.1   Planning Development Work 10
        2.3.2   Information Flow 10
        2.3.3   Feasibility Study During Product Planning 12
        2.3.4   Task Definition and Conceptual Stage 12
        2.3.5   Functional Specification 14
        2.3.6   Scheduling 15
    2.4   Technical Drawings 17
    2.5   Circuit Diagrams 22
    2.6   Computer-Aided Design (CAD) 24
    References 29

3   **System Architecture and Protection Requirements** 31
    3.1   Introduction—Terminology, Functions and Structures 31
        3.1.1   System Characteristics of Devices 32
        3.1.2   System Environment 32
        3.1.3   System Functions 33
        3.1.4   System Structure 34
    3.2   System Design Architecture 35
        3.2.1   System Granularity 35
        3.2.2   System Assembly 36
        3.2.3   System Integration in Environment 38
    3.3   Electronic System Levels 38
    3.4   System Protection 39
        3.4.1   CE Designation 40

       3.4.2   Protection Classes   40

       3.4.3   IP Codes of Enclosures   42

   References.   43

4 Reliability Analysis   45

   4.1   Introduction   45

   4.2   Calculation Principles.   47

       4.2.1   Probability Terminology   47

       4.2.2   Reliability Terminology.   49

       4.2.3   Reliability Parameters   49

   4.3   Exponential Distribution.   53

       4.3.1   Reliability Distributions.   53

       4.3.2   Reliability Parameters and the Exponential

               Distribution   55

   4.4   Failure of Electronic Components.   56

       4.4.1   Drift   57

       4.4.2   Reference and Operating Conditions   57

       4.4.3   Failure Rates of Electronic Components   58

       4.4.4   Derating   60

       4.4.5   Accuracy of Failure Rates   60

   4.5   Failure of Electronic Systems   62

       4.5.1   Calculation Principles   62

       4.5.2   Network Modeling—Serial and Parallel Systems   63

   4.6   Reliability Analysis of Electronic Systems   64

       4.6.1   Preliminaries   64

       4.6.2   Availability of Repairable Systems   65

       4.6.3   Electronic Systems Without Redundancy—Serial

               Systems.   66

       4.6.4   Electronic Systems With Redundancy—Parallel

               Systems.   68

       4.6.5   Service and Maintenance of Electronic Systems   71

   4.7   Recommendations for Improving Reliability of Electronic

       Systems   72

   References.   73

5 Thermal Management and Cooling   75

   5.1   Introduction—Terminology, Temperatures, and Power

       Dissipation   76

       5.1.1   Problem Definition   76

       5.1.2   Important Parameters in Thermal Management   79

       5.1.3   Temperatures of Components and Systems   82

       5.1.4   Power Dissipation in Electronic Components   83

   5.2   Calculation Principles.   84

       5.2.1   Electrical and Thermal Networks.   84

        5.2.2   Thermal Network Method   86
   5.3 Heat Transfer   90
        5.3.1   Introduction   90
        5.3.2   Conduction Heat Transfer   90
        5.3.3   Convection Heat Transfer   93
        5.3.4   Radiation Heat Transfer.   98
   5.4 Methods to Increase Heat Transfer   107
        5.4.1   Heat Sinks.   107
        5.4.2   Thermal Interface Materials.   110
        5.4.3   Fans.   110
        5.4.4   Heat Pipes.   112
        5.4.5   Peltier Elements.   113
   5.5 Application Examples in Electronic Systems   115
        5.5.1   Component Temperatures   115
        5.5.2   Outside and Inside Surface Temperatures of an
                 Enclosure.   117
        5.5.3   Choosing Open or Sealed Enclosures   119
        5.5.4   Heat Dissipation from Open Enclosures   121
        5.5.5   Heat Dissipation from Sealed Enclosures   124
        5.5.6   Heat Transfer Through Enclosure Panels.   130
        5.5.7   Interior Air Temperatures   134
        5.5.8   Heat Transfer Inside an Open Enclosure   135
        5.5.9   Heat Transfer Inside a Sealed Enclosure   137
        5.5.10   Forced Convection with Fans and Fan Selection.   138
   5.6 Recommendations for Thermal Management of Electronic
        Systems   145
   References.   146

6 Electromagnetic Compatibility (EMC).   147
   6.1 Introduction   148
   6.2 Coupling Between System Components   148
        6.2.1   Conductive Coupling.   150
        6.2.2   Capacitive Coupling   152
        6.2.3   Inductive Coupling   154
        6.2.4   Electromagnetic Coupling   156
   6.3 Grounding Electronic Systems   157
        6.3.1   Description of Reference Grounds.   157
        6.3.2   Reference Systems Schemes (Grounding Systems)   159
        6.3.3   Return Conductor to the Reference Point for Digital
                 Signals.   162
        6.3.4   Return Conductor to the Reference Point for Analog
                 Signals.   163
        6.3.5   Ground Loops   164

6.4   Shielding from Fields   165
      6.4.1   Shielding Fundamentals.   165
      6.4.2   Shielding Magnetostatic Fields   168
      6.4.3   Shielding Magnetoquasistatic Fields   170
      6.4.4   Shielding Electrostatic Fields.   173
      6.4.5   Shielding Electroquasistatic Fields.   175
      6.4.6   Shielding Electromagnetic Fields.   176
6.5   Electrostatic Discharge (ESD).   181
      6.5.1   Causes of ESD   181
      6.5.2   ESD-Suppression Measures.   182
6.6   Recommendations for EMC-Compliant Systems Design   183
      6.6.1   Key Steps in System Development   183
      6.6.2   Designing Printed Circuit Boards and Shielding   184
      6.6.3   Designing System Cabinets.   188
      6.6.4   Connecting Peripherals   190
References.   191

7 **Recycling Requirements and Design for Environmental
   Compliance**   193
7.1   Introduction—Motivation and the Circular Economy   194
7.2   Manufacture, Use, and Disposal of Electronic Systems in the
      Circular Economy.   197
7.3   Product Recycling in the Disposal Process.   199
      7.3.1   New Marketing Strategy—Selling Usage   201
      7.3.2   New Design Strategy—Product Durability   201
7.4   Material Recycling in the Disposal Process   203
7.5   Design and Development for Disassembly   206
      7.5.1   Structural Correctness   206
      7.5.2   Design for Disassembly.   208
      7.5.3   Ease of Opening   209
7.6   Material Suitability in Design and Development.   209
      7.6.1   Suitability of Quantities.   210
      7.6.2   Suitability for Separation.   210
      7.6.3   Suitability for Recovery.   211
      7.6.4   Material Compatibility.   213
      7.6.5   Suitability for Disposal   213
      7.6.6   Material Labeling   215
7.7   Recommendations for Environmentally Compliant Systems   215
References.   217

8 **Appendix**   219
8.1   Notes and Rules on Technical Drawings   219
      8.1.1   Title Block   219
      8.1.2   Scales   220

	8.1.3	Identification Number	220
	8.1.4	Paper Sizes	221
	8.1.5	Line Styles and Widths	221
	8.1.6	Sectional Views	222
8.2	Geometric Dimensioning and Tolerancing		224
	8.2.1	Elements of Specified Dimensions	224
	8.2.2	Dimension Types	225
	8.2.3	Tolerance Terminology	225
	8.2.4	Engineering Tolerances	226
	8.2.5	General Tolerances	226
	8.2.6	ISO Tolerances	227
	8.2.7	Form and Positional Tolerances	227
	8.2.8	Surface Specifications	227
	8.2.9	Material Specifications	228
8.3	Preferred Numbers—Renard and E-Series		228
8.4	Schematic Symbols of Electronic Components		231
8.5	Labeling of Electronic Components		234
	8.5.1	Labeling with Colors	234
	8.5.2	Labeling with Characters	235

Index 237

Chapter 1
Introduction

Electronic systems design is the subject within electrical engineering that deals with the multidisciplinary design issues of complex electronic devices, such as smartphones and computers. The subject covers a broad spectrum, from the development of an electronic system to assuring its proper function, service life, and disposal. Major advances in technology, the increasing multidisciplinary nature of the development process and the use of electronic devices in all aspects of our daily lives pose immense challenges for every design engineer.

The book covers all aspects of the development of electronic systems by presenting the theoretical knowledge required for their design and fabrication. This is a discipline that spans electronics, physics, mechanics, and other topics. Designers of electronic circuits, on the one hand, often lack the necessary manufacturing and overall system's expertise, while, on the other hand, (electro-) mechanical designers are hindered in their work by their lack of knowledge of electronic components. This is where this book comes in; it aims to marry the various disciplines involved.

The goal is to convey the knowledge and skills necessary for designing and developing electronic systems and an understanding of the myriad engineering approaches and tasks involved. The reader should learn from the book how to work as a designer and fabricator of these products and acquire the necessary knowledge of all relevant aspects. The key issues encountered in the development of electronic systems are pictured in Fig. 1.1 along with references to the respective chapters in the book.

The principle topics covered are the design process, packaging issues, and associated system levels, extended with special requirements for the development and fabrication of an electronic system. These requirements include protection issues, reliability, thermal management and cooling, shielding, and recyclability. The layout of the book is detailed below:

Chapter 2, *Design Process and its Fundamentals,* presents the steps involved in the design process for electronic systems as well as the use of technical design documentation, such as technical drawings and circuit diagrams. It also provides an introduction to computer-aided design (CAD).

© Springer International Publishing AG 2017
J. Lienig and H. Bruemmer, *Fundamentals of Electronic Systems Design,*
DOI 10.1007/978-3-319-55840-0_1

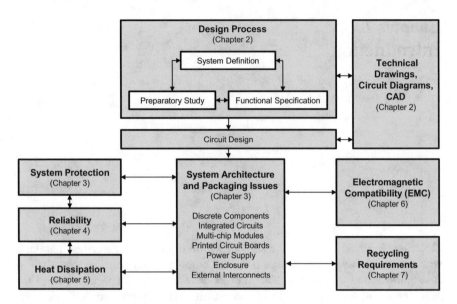

Fig. 1.1 Requirements for the development of an electronic system and the matching book structure

Different packaging methods for the system-level and for individual components as well as system protection are described in Chap. 3, *System Architecture and Protection Requirements*. Particular emphasis is put on protection classes and IP codes which stipulate how a system should be designed with the protection of persons and the device interior in mind.

Critical reliability parameters and their use are introduced in Chap. 4, *Reliability Analysis*. The reliability requirements for system-level and package design can thus be met and the overall reliability of a system calculated from the known reliabilities of individual components.

Losses and heat transfers associated with components and the overall system are covered in Chap. 5, *Thermal Management and Cooling*. Thermal characteristics can be determined at the design stage and suitable elements selected and deployed for heat dissipation and meeting thermal criteria.

Chapter 6 in the book, *Electromagnetic Compatibility*, deals with EMC issues when designing electronic systems. It also covers conceptual solutions comprising shielding and protection measures against electrostatic discharge (ESD).

Chapter 7, *Recycling Requirements and Design for Environmental Compliance*, presents material that may be new for many engineers and will certainly increase in importance as industry continues to evolve. The chapter describes critical environmental considerations during the design and development stages that have tremendous impact later in the product life cycle, in particular at the tail end during product disposal and recycling. The challenge for designers in today's waste-disposal-aware society is to produce environmentally compliant systems. Waste and energy consumption should be minimized during manufacture, use, and disposal, and system materials should be fully recyclable after use.

The appendix (Chap. 8) presents rules of technical drawings, preferred numbers (Renard and E-Series) and schematic symbols for electronic components, including their labeling with colors and characters.

Chapter 2
Design Process and Its Fundamentals

In this chapter, we describe the basics of the development process for electronic systems. We will see how service-proven standards and norms along with standard drawings and computer technology can be used to break down the design process into separate activities, which are then more easily performed. Every research and development engineer needs to be familiar with these design activities and with the requisite technical documentation (technical drawings, circuit diagrams, CAD models) in order to produce successful electronic products.

2.1 Life Cycle of Electronic Products

The electronics industry is an exciting place to be these days, and indeed most technical innovations today come from this industry. Fifty percent of many companies' sales come from products that are less than five-years old. These products need close attention not only just when they are first designed but also through their entire life cycle. Figure 2.1 shows the typical life cycle of a product from a business standpoint. The product life span can vary. In order for a product to be economically viable, a business must establish whether the development costs can be recouped and, if yes, over what time period.

The different stages in a product's life, also known as the product life cycle, typically consist of development (*development stage*), use (*marketing stage*) and disposal (covered in Chap. 7). The development stage consists of the following steps:

- product planning,
- design and development, and
- first production run with prototype build and pilot series.

As we can see in the middle of Fig. 2.1, the growth rate is high at the beginning of the marketing stage and then the product matures. The market for the product and

© Springer International Publishing AG 2017

J. Lienig and H. Bruemmer, *Fundamentals of Electronic Systems Design*,
DOI 10.1007/978-3-319-55840-0_2

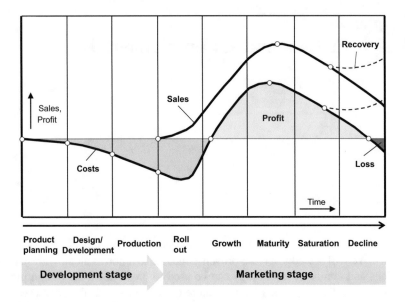

Fig. 2.1 Development and marketing stages for a product [1, 2]

the competition are known. The maturity stage is superseded by the saturation stage, where there is only low or no growth. As more players enter the market, the sales price comes under increasing pressure. Direct costs are driven down by increased productivity. But, despite this, the relationship between sale price and direct costs is increasingly reduced. Finally, during a period of decline, the product is typically forced from the market by the competition or by the other substitute products.

The development process for an electronic system described in this book begins with the *product planning* stage. During this initial planning period, ideas for products are generated and assessed, and project tasks are formulated. The subsequent *design and development* stage uses as input the functional specification developed with the product proposal. At the end of the design and development stage, the complete product documentation containing all instructions for product fabrication, use, maintenance, and disposal (including recycling) is produced.

2.2   Design and Development Process

The design and development of an electronic system or device is a key element of the preparations for manufacture; in essence, function, quality, and costs are defined at this stage. *The design process encompasses all creative, manual, and technical activities necessary to define the product and which need to be carried out to convert a system definition to a sufficiently detailed system design specification for product manufacture and deployment.*

Before we proceed, it is helpful to clarify a bit of terminology. The term "design" typically covers the first prototype for a system or a device, while the term "development" generally includes the production of the documentation (e.g., circuit diagrams for electronic modules, drawings for mechanical components) necessary for fabrication. As it is often difficult to differentiate these two terms, they are considered as synonyms throughout this book.

Design and development can be divided into four stages, each with different definitions [1]:

- task definition (informative definition),
- conceptual stage (cardinal definition),
- design stage (formative definition), and
- implementation stage (manufacturing definition).

Information is gathered on the requirements for the product to be developed during *task definition*. The result is an informative definition in the form of a requirements specification. These requirements are then transformed into an optimal technical principle at the *conceptual stage*. The cardinal definition of a solution is thus formulated here along with proof of functionality.

A system design based on the conceptual solution is presented at the *design stage*. The objective is to determine the best overall design for the product taking technological and economic constraints into consideration. Finally, fabrication and usage details are set out at the *implementation stage* to facilitate manufacturing and deployment.

Engineering guidelines usually define seven steps for the design and development process, which add further details to the four stages above:

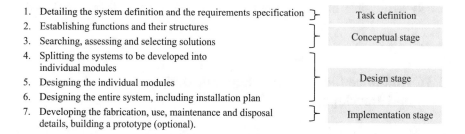

Figure 2.2 shows the steps involved in the overall product design process. A number of variants, which subsequently need to be optimized and streamlined, are produced in several of the steps. Design teams often use creativity techniques, such as brainstorming, to produce these variants. Variants are optimized and eliminated with the help of selection and assessment methods, such as weighted lists [1].

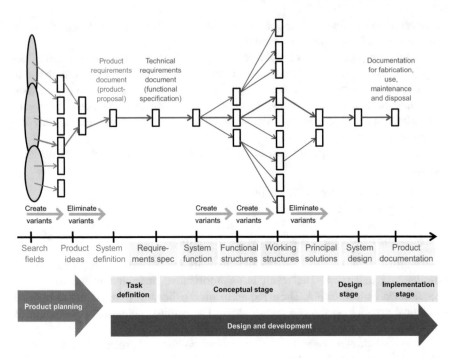

Fig. 2.2 Steps and results of the product planning and development processes. Development effectively begins with the system definition based on the product proposal. The development work is divided into the task definition and the conceptual, design and implementation stages. How variants are produced and eliminated is also shown

At the end of the design and development process, a prototype is typically produced. This first prototype should be tested under conditions that replicate as closely as possible the conditions that will be encountered later in real-world scenarios.

2.3 Guidance for Product Planning, Design and Development

In today's consumer-driven industry, development departments must respond rapidly to market changes. They need to draw up a work schedule along with cost estimates to ensure optimal use of resources. This is particularly critical for more complex systems, where knowledgeable specialists from a wide range of backgrounds are often involved. As a prime example of a complex system, we shall examine a "vending machine" whose block diagram is shown in Fig. 2.3. The vending machine example will illustrate key concepts and steps that design teams will use for a broad range of consumer and industrial products.

Fig. 2.3 Block diagram of a vending machine

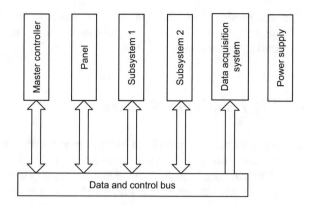

In our vending machine example, the user selects an item via a panel with keyboard and screen which also displays the price. Coins are identified and their values determined by the cashier unit (subsystem 1). The coins are stored in the device. The machine gives change if needed on a command from the master controller.

The development of an electronic coin validator with high detection reliability for counterfeit coins and other currencies requires a team of qualified electronic, mechanical, and test engineers, drawing upon their specialized knowledge. The system definition needs to be precisely formulated; duties and responsibilities should be clearly set out so that there is very little wasted capacity. The same applies to the merchandise storage space (subsystem 2) and associated output unit with actuator.

The master controller checks all operations in the vending machine. When a button is pressed and an item is selected, the price is read from an electronic price register and displayed. When payment is received, the controller calculates the difference between the sale price and the value of the inserted coins and displays it on the screen. If change needs to be given to the customer, the cashier unit returns the largest coins possible. The master controller controls all drives in the vending machine, calculates the values for data acquisition, and outputs an alarm message in the event of a fault in a module.

The data acquisition system stores sales transactions over a longer period and gathers statistics. The power supply unit powers circuits, screens, and actuators.

All functions and subfunctions must be accurately defined when designing such a system, to ensure correct operation. Even in this introductory example of the vending machine, we can see how complexity grows as our requirements become more detailed and specific. We manage this complexity by defining high-level functions, subfunctions, and interactions with which we will implement these requirements. The specification also depends on the type of components and circuits to be used; for example, conventional digital logic or a computer could be deployed as master controller. The project engineer is therefore required to investigate and define the system early in the project. As the number of people involved in the project grows, so too do costs and coordination issues. It is therefore imperative from a business perspective that joint activities within the team be effectively scheduled.

2.3.1 Planning Development Work

The argument is often made that development work cannot be planned. However, this is not correct, as the truly creative work only takes on average 10–15% of the total project time, with the remaining time taken up with well-defined tasks that will transform these creative ideas into a viable product. Workflows and planned times for such tasks should be defined by the engineers involved in the project, as they are the only people with the necessary knowledge.

First, as much information as possible should be collected on the project. The mission should then be formulated. Subsequently, different solutions will be developed and evaluated. Once the best solution based on the requirements specification has been selected, the system to be developed is defined, the engineering work is set out, and projected costs are calculated.

The timetable for the design process should not be seen as a restriction on the work of the development engineers or as pressure on them. Rather, its purpose is to make the complex decision process more transparent and get superiors involved. Indeed, management has to take responsibility for decisions. More often than not they have a better understanding of how the overall system fits in the marketplace and of customer concerns, but have less detailed system knowledge.

Project planning increases the likelihood of success by preventing false starts; it permits the development engineer to start a project only after all solutions have been examined and assessed; and the design criteria established as per the requirements specification. The project engineer should study all technical details before starting any development work.

He/she should document all his/her reviews and investigations as well. This is very important, so as to allow others to participate in decision making for certain solutions or for the entire project, and to understand later why and how certain decisions have been made. In addition, as new engineers join the team, they can quickly come up to speed by reviewing such documentation and understanding key decisions.

2.3.2 Information Flow

The development engineers should work closely and continuously with the customer right from the start. The customer is the company's customer in the case of stand-alone, customized devices. Sales, i.e., the marketing department, often assumes the role of the customer for standard, mass-produced products. Typically, they know the problem to be solved, but not the deployed technology. The opposite is true for development engineers. Hence, unnecessary delays and costs will be avoided when there are competent contact persons on both sides, working closely together in an iterative, collaborative manner.

Knowledge of certain merchandise groups and understanding the requirements restricting the system to these merchandise groups is pivotal in the case of the vending machine example introduced earlier. For example, overall costs can likely be significantly reduced if very large or heat-sensitive items of merchandise are not offered by the machine. Customer options to adjust the price register, the statistical data gathered, the UX/UI (User Experience/User Interface), and the system design are other topics that must be considered.

Direct contact between customer and the company's development engineers will facilitate a speedy and seamless system handover. Costs for the proposed solutions should only come from sales people as they are the only people who know the valid cost plan. The development team should be in close contact with the fabrication department, as the manufacturing technology used in a company typically has a major impact on product design.

Companies are often organized as per Fig. 2.4. The term "development" often stands for the development department with its own R & D laboratory and design departments and may be used to describe research laboratories as well.

All electrical, electronic, physical and chemical investigations, and experiments needed for system development are typically carried out in laboratories in many companies. Circuit diagrams and documentation, used, for example, for IC and

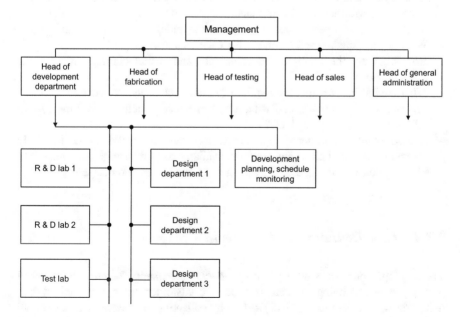

Fig. 2.4 Organization chart of a company

PCB layout design and packaging, and technical drawings for mechanical components are produced during these tests. Prototypes are also assembled and tested in test areas at this time.

2.3.3   Feasibility Study During Product Planning

The system to be developed is first roughly described in a *feasibility study* during product planning, in preparation for system development. It may also be possible at this stage to determine the features a system should not have if system characteristics cannot yet be clearly defined. The feasibility study typically contains details on the application and the customers, an analysis of competitor systems (rival products), and a rough estimate of the sales targets and required total investments.

The purpose of the feasibility study is to help the company decide whether or not to go ahead with the project. The feasibility study need not be carried out if the company executives decide to go ahead with the project for other reasons, such as marketing.

A typical feasibility study for a vending machine, for example, would address the following questions:

- What customers and what merchandise groups should the system be designed for?
- What characteristics does the system need to have?
- Is the system an economically viable proposition for the customer considering price and performance? Has the customer expressed maximums or minimums for price or performance, respectively, and/or trade-offs between them?
- What is the expected sales volume for each customer group?
- How high are the maximum development and manufacturing costs for the assumed price?
- What capital investments are required to manufacture the system?
- What is the competition, and what advantages and benefits should the system offer over competing products?
- Are some of the system components, like the coin validator or the fully assembled cashier unit in the example, available as purchase items? This might be an option to eliminate development risks or cut development costs.

2.3.4   Task Definition and Conceptual Stage

The development process starts when a *product requirements document* is produced during product planning or there is a definite customer order. First, the system definition will be detailed (*task definition*) by collecting all necessary data from all available sources. Among the topics covered are: establishing the state of the art of the given technology, standards and norms, legal issues, the protection of utility models and patents, and any necessary authorizations.

The task definition reduces the level of abstraction for the design solution, and it becomes more concrete. At the same time, it defines the functional specifications for the development process.

The task definition produces the *functional specification* based on the customer *product requirement document* (Sect. 2.3.5). This *functional specification* thus contains, from the developers' view point, the company's specifications for building the system.

An intermediate report should be produced after the task definition. A decision will be taken about continuing the project based on this report. Solution options will then be identified for the proposed system, and standards for selecting the optimal solution will be formulated as well. The following tasks will then be carried out:

– Drafting the definitive functional specification (Sect. 2.3.5).
– Further detailing of the selected solution for identifying any major problems and drawing up a project structure plan (Sect. 2.3.6).
– Defining the reliability data and the maintenance intervals.
– Performing trials for components that have not been proven yet. These tests may continue until the end of the development period. Alternative solutions may be needed.
– Drawing up a training program for service personnel and producing service documentation for the course.
– Determining criteria (e.g., functions, costs, and time) whose compliance is required for securing the project. These criteria should be reviewed regularly during the project.
– Drawing up a development time schedule (Sect. 2.3.6) and a report with recommendation (or disapproval) to continue with the project.

The design stage should only start if the solutions submitted after the conceptual stage have been successfully verified. Junior engineers often make the mistake of ending the conceptual stage too soon, i.e., they start designing components too quickly before all solutions have been identified and assessed.

The project work described here is often affected by modifications to the system definition after the functional specification has been drafted and before the system is deployed. These changes could be caused by market changes, by rivals, or by other technical requirements from the customer. The impact of such changes on delivery dates and costs can be more accurately estimated if sufficient design and engineering documents are available when such a situation occurs.

Executing the project as described above brings transparency to planning and decision making for higher management, and increases the chances of technically and commercially successful project completion.

2.3.5 Functional Specification

The design and development process starts with a functional specification, typically outlined in a *technical requirements document*. This is a list of requirements the system has to comply with, along with the technical definitions of the operating environment. These requirements are the technical response to a matching requirements document, the *product requirements document*, created from a user's point of view by the contracting party or the product planning department. Hence, the functional specification contains engineering details for developing the system based on the product requirements document from the customer.

Requirements should be ordered according to their priority:

- Requirements that the system must fulfill for the targeted customers,
- Features that could attract a wider group of customers if the price is not increased at all or is only marginally increased, and
- Low-priority requirements that do not necessarily have to be complied with ("nice to have").

A typical functional specification will contain the following:

- Precise description of the system definition (what is the purpose of the product or system?),
- Description of the interfaces to the environment, such as other technical systems and humans,
- Size and weight definitions and installation conditions,
- Definition of the operational and environmental conditions,
- Standards of precision,
- Functional safety,
- Service life duration,
- Maintenance and repair requirements,
- Standards and regulations, such as mandatory standards,
- Storage conditions, transportation requirements, and packaging,
- Sales volumes,
- Approved development and manufacturing costs, sales price, running costs at customers, and
- Deadlines.

By asking product-specific questions, such as "Does the customer have experience with similar devices or components?" "What characteristics should the product *not* have?", additional information is often obtained that should be defined *a priori*.

2.3.6   *Scheduling*

Scheduling the design process can be a major challenge. There is no question that it is needed; management rightly demands specification and compliance with deadlines and costs from engineering departments. Proper scheduling helps avoid frustration and costly delays, which can adversely affect both management and engineering "enthusiasm" for the project. As the saying goes, *plan the work, then work the plan*.

A number of requirements must be fulfilled to create a useful planning system. The activities of different employees and departments in product development must be coordinated and scheduled. A *network plan* is particularly useful in this regard for large projects. This type of time schedule shows all project activities and their mutual dependencies in graphic form, as will be explained later. The following conditions need to exist for a successful network plan:

- The planning system must match the organizational structure;
- A competent member of staff should be responsible for planning and scheduling;
- All employees should be familiar with the planning system and should endorse it; and
- Sufficient time is allocated for the preparatory study and planning.

There is a difference between the planning system and its data. A lot of publications are available on different planning systems and techniques [1, 3].

A *project structure plan* (Fig. 2.5) is typically drafted when preparing a proposed development solution. This is a simple and self-explanatory schematic

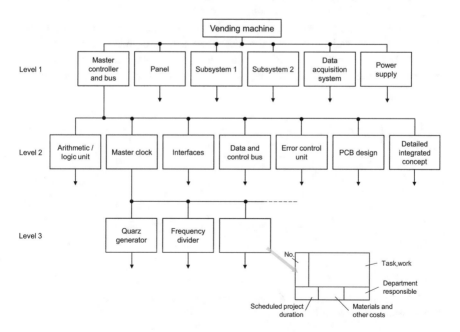

Fig. 2.5 Section of a project structure plan for the vending machine depicted in Fig. 2.3

representation of the work needed to complete the project. The individual levels in the project structure plan indicate the different subtasks and their relationships to higher- and lower-level tasks. The subtasks, representing the final items of the schedule, are called *work packages*.

The project structure plan is not yet the network plan. The former should always be drafted to ensure that no critical items are overlooked in the latter.

The work packages should be small enough to allow accurate *time estimates* for their completion. Practicable time units, such as hours, working days or weeks, should be used. The more detailed the work description, the more accurate the scheduling. And the more detailed the information available on the tasks, the better the resulting plan will be. One of the purposes of the preparatory study is to examine the less clearly defined topics and to gather as much information as possible on them. The work will need to be divided into smaller packages if some issues remain unresolved. A review will then be carried out at the end of each work period, and work rescheduling may be necessary.

When estimating times, remember that development engineers are often sidelined from their design work such as they may spend up to 40% of their time at meetings, talking on the phone, writing e-mails, and doing other works.

When the project structure plan has been drawn up, the work packages are divided into *processes*. A process is a time-consuming activity with a defined start and finish. Another process attribute is that it incurs costs. Additionally, a process is performed without a break from start to finish.

Individual processes in a project cannot be performed in any order. It may be necessary to start a given process only when other processes are complete. This gives rise to *relationships* and *dependencies*.

The logical project plan is defined and the *network plan* can be drawn up when the predecessors and successors for individual processes have been established (Fig. 2.6). It is sometimes easier to draw up the network plan starting with the project end date and "work backwards." This is because it is often simpler to

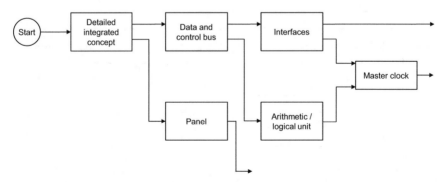

Fig. 2.6 Section of a network plan (activity-on-node network) based on the project structure plan for the vending machine in Fig. 2.5

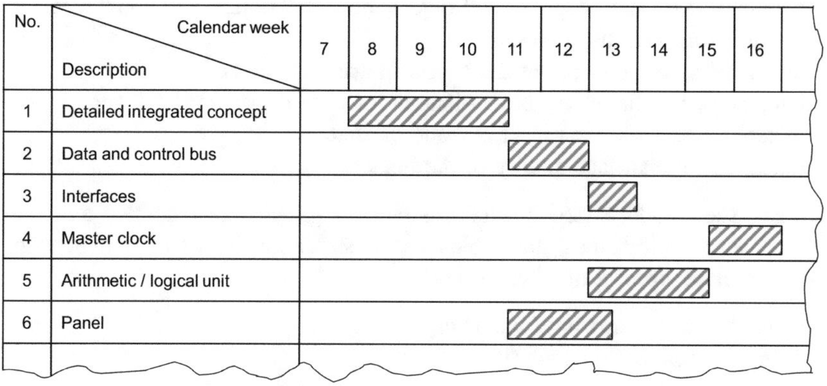

No.	Description	7	8	9	10	11	12	13	14	15	16
1	Detailed integrated concept	▨	▨	▨							
2	Data and control bus					▨	▨				
3	Interfaces							▨			
4	Master clock									▨	▨
5	Arithmetic / logical unit						▨	▨	▨		
6	Panel					▨	▨				

(Calendar week)

Fig. 2.7 Section of a bar chart

establish what needs to be done before a process is started than to determine the steps needed after its completion.

The earliest and latest start and end times for individual processes are calculated from the relationships and time estimates. There will be many free periods (buffer periods) available for many processes. The path from the start to the end of the network plan, where there are no buffer periods, is called a *critical path*.

Finally, the time units used for planning are converted to calendar dates to build a *bar chart* or *Gantt chart* (Fig. 2.7). The following must be known for calendar planning:

- the project start date (calendar date),
- holidays during the project period, and
- the planning unit (days, weeks).

A bar chart can be produced without creating a network plan with projects that have very few processes and very few mutual dependencies.

2.4 Technical Drawings

All parts have to be uniquely and adequately described for fabrication in the *product documentation* generated at the end of the design and development process [4]. Freehand sketches of components, modules, and systems in compliance with standards should be read and produced at the conceptual stage. Typically, development engineers are required to deliver complex, often computer-based, *technical drawings*.

In order to communicate all needed information from the development to the fabrication process, a technical drawing must include the following critical information:

– geometry, i.e., the shape of the object, how the object will look when it is viewed from various angles,
– dimensions, i.e., the size of the object in accepted units,
– tolerances, i.e., the allowable variations for each dimension,
– material, i.e., what the object is made of, and
– finish, i.e., the surface quality of the object.

Using such standardized illustrations ensures global understanding and portability of the specifications and avoids misconstructions. A set of drawings for a system is made up of a number of items:

– main or general drawing (mandatory),
– group or module drawings (if required),
– detail part drawings (mandatory for all parts to be manufactured),
– bill of components (contains all system parts, also standard or purchased parts; it is mandatory), and
– assembly drawing (optional).

The *right-angled parallel projection* as per ISO 128-30 [5] has established itself as the benchmark for parts drawings. It has the advantage that all views can be displayed undistorted and true to scale with all dimensions. The front and side views are produced by the *orthographic projection*, which is the standard technique for drawing a physical object in different plan views (Fig. 2.8).

An object can have six views, whose defined location and orientation to one another are derived from the orthographic projection (Fig. 2.9):

	Bottom view		
Side view from the right	Front view	Side view from the left	Rear view
	Top view		

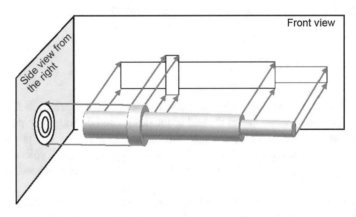

Fig. 2.8 Orthographic projection of an object for representing the front and side views. All visible object edges and the silhouette are shown in the viewing angle for each view

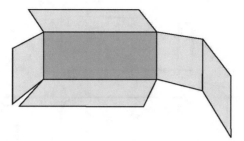

Fig. 2.9 Six views of an object are generated by projecting in the respective planes of a rectangular expandable projection box. The side view from the right, for example, is located on the *left* of the front view (see Fig. 2.8), and the bottom view is located *above* it

For engineering documentation, the views drawn are typically a subset of all six, namely only the ones needed to uniquely render the object. As an example, Fig. 2.10 pictures an object in all six views and Fig. 2.11 shows only the necessary views to be drawn.

The size of a drawn object when it is increased or reduced in size with respect to the original is indicated with a *scale*. While a scale of 1:1 represents the original size, enlargements may be represented by scales of 2:1 (20:1, 200:1, etc.), 5:1 (50:1, 500:1, etc.) and 10:1 (100:1, 1000:1, etc.). The same applies to size reductions that

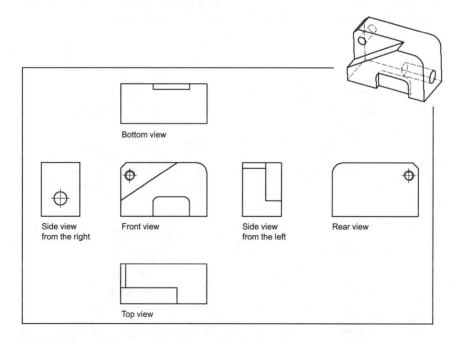

Fig. 2.10 Six possible views of an object and their orientation to one another

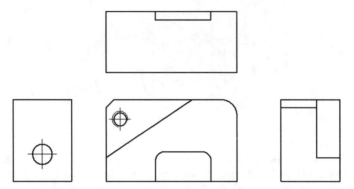

Fig. 2.11 Object in Fig. 2.10 can be uniquely shown in four views (front view, side views from left and right, bottom view)

can be drawn in the scales 1:2 (1:20, 1:200, etc.), 1:5 (1:50, 1:500, etc.) and 1:10 (1:100, 1:1000, etc.) [6]. The scale is specified in the title block on the drawing [7].

Sectional views, also known as sections, are used to illustrate features inside an object or to reduce the number of views (Fig. 2.12) [8]. Cut surfaces are shaded, and cut surfaces for the same part are shaded in the same way. Every sectional view is based on a non-sectional view of an object. The location of the sectional view on the unsectioned object may be indicated by a dash-dotted line for clarity purposes. The viewing direction is marked with arrows. Note that complete workpieces, such as shafts, pins, and screws, remain unsectioned in a sectional view [8].

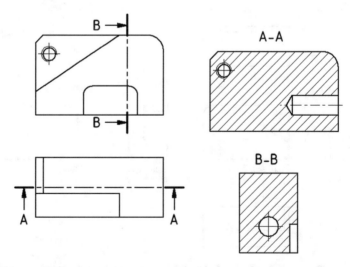

Fig. 2.12 Sectional views from front and top views for the object in Fig. 2.10 illustrate the locations and orientations of the holes, in particular

If the depicted object is drawn for manufacturing or testing, it must be uniquely *dimensioned* (Fig. 2.13). There are different types of dimensioning: function-based (functional dimension, i.e., the dimension is essential to the function of the piece or space), production-related (the dimension is relevant for manufacturing), and test-related dimensioning (inspection dimension). A functional dimension is specified only when needed. Functional dimensions should be specified first in a

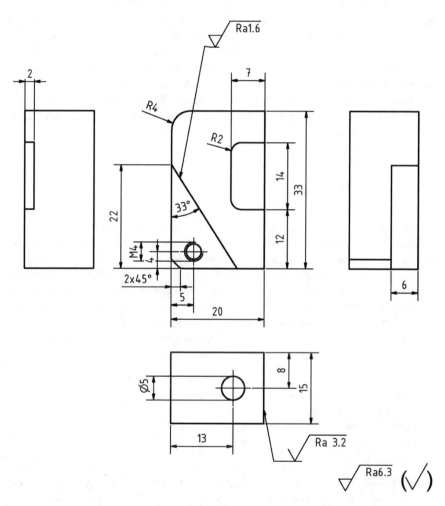

Fig. 2.13 Object in Fig. 2.10 with production-related dimensioning. There is no need to specify tolerance for the individual dimensions as general tolerances are specified in the title block ("ISO 2768–mK", not shown in the figure here). Solid surface specifications that specify an average surface roughness of 6.3 μm to be achieved by material abrasion are given *bottom right*. The bracketed expression indicates that there are also surfaces with different tolerances (individually labeled in the drawing)

drawing, followed by the production-related dimensions and the inspection dimensions. Inspection dimensions are typically used in inspection drawings only.

Every dimension should be quoted with a *tolerance value* as no part can be perfectly manufactured. General tolerances, which apply to all dimensions in the drawing and which are specified in the title block, should be preferred (see Fig. 2.13). Other options include (1) the ISO system of limits and fits which is a worldwide coordinated system of hole and shaft tolerances, and (2) manually specifying tolerance limits (see Chap. 8.2).

In recent decades, technical drawings have increasingly been prepared using computer technology with its CAD systems. Software programs for 2D drawings that provide different views of an object were initially developed, which essentially replaced paper-based drawings with digital ones. Today's CAD packages add new functionality to technical drawings; not only can objects be represented in 3D, they can also be optimized with comprehensive 3D modeling (Sect. 2.6).

Chapters 8.1 and 8.2 contain the most important instructions and guidelines for technical drawing; Chap. 8.3 contains preferred numbers and dimensions to be used in the design process.

2.5   Circuit Diagrams

Modern systems contain mechanical and electronic components. Circuit elements and their interactions must be represented in *circuit diagrams*, also known as *circuit schematics*. Development and design engineers need to understand circuit diagrams and know how to produce them. A circuit diagram is a symbolic description of a circuit showing the circuit elements and the links between them. The layout of the components in the diagram is not the same as the position of the elements and the connections in the actual implementation. The main purpose of the circuit diagram is to show the logical and electrical functions and interconnections of the circuit. It contains the following elements (Figs. 2.14 and 2.15):

– symbols
– device labels (ID letter with consecutive numbers),
– device type or rating,
– electrical connections (interconnects, bus systems),
– back-annotation data (optional), and
– frame and title block.

The *symbols* found in a circuit diagram can be either elementary symbols for resistors and other component symbols, or block symbols, e.g., subcircuits.

Each symbol has a unique *label* composed of a letter, which defines the component type, followed by a serial no. (*C4*—capacitor no. 4, *D1*—digital gate no. 1, *R12*—resistor no. 12). The component *type* (e.g., gate *74ACT04D*) or *value* (e.g., resistance 10 k) should be written beside the symbol as well. The engineering unit

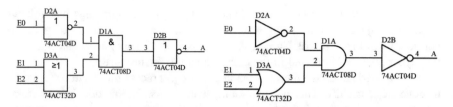

Fig. 2.14 Section of a circuit diagram with *digital* components (AND gates, OR gates, two inverters) in IEEE/ANSI/IEC standard format (*left*) and traditional schematics format (on the *right*). The letters A or B added to the labels *D1, D2,* and *D3* indicate copies (instances) of a library element. Elements *D2A* and *D2B* are copies A and B of inverter *74ACT04D* in the library in this example. Both gates are contained within the chip package *D2*

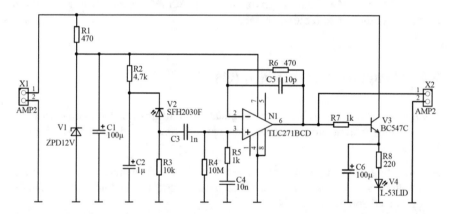

Fig. 2.15 Exemplary circuit diagram (with no frame or title block) with an operational amplifier and different *analog* components (resistors, capacitors, connectors, photodiode, Zener diode, LED, transistor). Each symbol is followed by a letter with a serial no. and the type and/or value of the component. The engineering unit (e.g., ohm or Ω) is not cited

(e.g., ohm or Ω) is not written with the values. *Pin numbers* are specified where necessary to avoid ambiguities (e.g., with integrated circuits, connectors) and when they are not already defined by the symbol (as with the transistor).

Digital elements are typically located in a common IC package despite being drawn individually in the circuit diagram. This assignment can be indicated in the schematic view. For example, *D2A* and *D2B* are two copies A and B of a type *74ACT04D* inverter, which are contained in package *D2* (Fig. 2.14 and Chap. 8.4).

The following *connections* are deployed in a circuit diagram:

– lines: of no electrical significance, for decorative purposes only, e.g., borders;
– wires: pin-to-pin interconnects of signal paths, signal name is optional; and
– bus systems: bundling many signal paths, signal name is obligatory, every signal has the same name and a different index.

Back-annotation data are details gathered during the design steps that follow circuit design, e.g., layout design, which you "write back" in the circuit diagram to consider them in the future. Current values for specific interconnects calculated in the layout could be inscribed in the circuit diagram, for instance.

A completed circuit diagram is used in the design process for electronic circuits to generate a net list. This net list, along with device information loaded from libraries and technology information, is then used to produce the implementation arrangement, the *layout*, for the integrated circuit (IC) or the printed circuit board (PCB).

Please refer to the end of the book (Chap. 8: Appendix) for more information on circuit diagrams and components. Nominal values for devices based on the pre-ferred numbers are defined in Chap. 8.3. Chapter 8.4 contains a list of symbols for devices for use in the circuit diagram, and Chap. 8.5 introduces labeling options of electronic components (colors, characters).

2.6   Computer-Aided Design (CAD)

Computer-aided design (CAD) is the application of information technology to aid in the creation, modification, analysis, and optimization of a design. CAD features include:

- producing and using computer models for different disciplines (e.g., mechanics, electronics, heat transfer, fluid mechanics, magnetics) to find suitable and/or optimized solutions;
- deriving direct production information (e.g., rapid prototyping, CNC (Computerized Numerical Control) fabrication); and
- generating product documentation (e.g., technical drawings, production docu-ments, manuals).

A *CAD model* is typically seen as a computer model of a device or a system that includes its geometric and material properties. That said, it can also reflect and optimize the different requirements that need to be considered in the design and development process of a device or a system (Fig. 2.16).

The CAD model represents the system design status at a specific time in the design process. The first version of a CAD model is typically produced at the end of the conceptual stage in the design process, modeling the geometric and material properties of the conceptual solution.

Additional numerical "special-purpose models" cover selected requirements based on the current state of the CAD model. Among these extra models are finite-element models for heat issues, structural mechanics, and electromagnetism as well as electrical and timing circuit models.

For many decades, computer-aided design has played a key role in the design of complex VLSI circuits with millions (and now, billions) of gates.

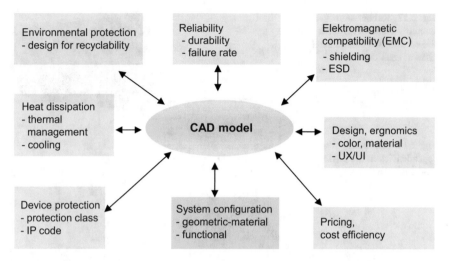

Fig. 2.16 CAD model at the heart of the system design process. Besides modeling the geometric, material and functional device/system configuration, other special-purpose models, such as finite-element models for optimizing heat dissipation, can be built for specific requirements in the design process

Geometric-material aspects of the electronic circuit are taken into consideration in the CAD model, e.g., as external interconnect and pad dimensions for the associated electronic devices. "Special-purpose models" form the basis for the following activities:

– producing a circuit diagram from a digital logic description,
– simulating and verifying the circuit function(s),
– automated design of the layout for the substrate (IC and PCB), and
– verifying the layout against the design constraints.

Simulation runs based on these "special-purpose models" are typically used to analyze the system for improvement with respect to individual requirements. The results of these analyses lead iteratively to an optimized CAD model for the system to be developed, whereby the geometric and material design characteristics are modified. CAD models thus assist the engineers in iteratively optimizing the technical solution throughout the design process.

We will show how a detailed analysis can be performed based on the geometric-material CAD model with an example of a simple mechanical rubber stop. The rubber stop consists of two carbon steel disks, with a rubber sleeve between them, acting as a resilient and damping element (Fig. 2.17).

The CAD model of this rubber-stop assembly is made up of the CAD models of two identical carbon steel disks and a rubber sleeve (Fig. 2.18). The relationship of the components to one another is described in the assembly model with parameterized assembly dependencies (e.g., "Insert" for parallel and concentric orientation of circular surfaces).

Fig. 2.17 Rubber-stop
assembly made of a rubber
sleeve with two carbon steel
disks at each end

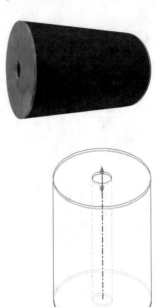

Fig. 2.18 Assembly
dependencies in the CAD
model of a rubber stop

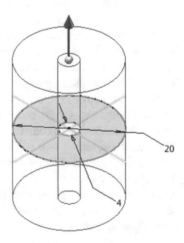

Suppliers usually provide CAD models for purchase parts. The CAD models are available in the libraries of CAD systems for standard parts (e.g., screws and nuts). CAD models need to be generated for custom-designed components only.

The user can gradually produce three-dimensional (3D) geometrical models of parts from two-dimensional (2D) sketches with geometrical operations (e.g., extrusion or rotation about an axis). The sketched contour of the circular cross section is extruded to create this rubber sleeve model (Fig. 2.19). The model could also be created by rotating a rectangular sketch about the z-axis.

Fig. 2.19 Producing the
3D-geometric model from 2D
sketches (volume of rubber
sleeve)

In addition to this example, CAD systems feature the following options:

– Complex geometrical shapes can be created from basic elements with Boolean operators (e.g., union, difference, average).
– Standard form elements (e.g., holes, screw threads, bevels) are supported by features in CAD systems. All you need to do then is place a form element (e.g., bore hole) at the desired position and configure it (e.g., with a screw thread or counterbore).
– Dimensions used in the geometrical operations and form elements in the 2D sketches can be modified at any time in the design process. All dependent dimensions will automatically change as well, including the component arrangements in the assembly, for example.

In addition to utilizing these capabilities for creative design, development engineers still also use CAD systems for generating sets of drawings from the 3D geometric models. The required dimensioned, scale views can be semi-automatically generated in the part drawings (Fig. 2.20).

So-called exploded views can be produced as well. They are especially suitable for highly complex structures at the component level to illustrate the assembly of parts and subassemblies. Figure 2.21 shows an example of the basic rubber-stop assembly with a semi-automatically generated parts list.

Besides using CAD systems to produce a set of drawings, they are also applied in *finite-element simulations* (*FE simulation*) to analyze the mechanical character-istics of components. The finite-element models are automatically generated based on the 3D-geometrical models. The example in Fig. 2.22 shows the vonMises equivalent stress distribution in the rubber sleeve when external compressive forces are applied only on the hole edges of the carbon steel disks. We can determine where the maximum stress occurs by means of mechanical finite-element simula-tion. The shape of the part can thus be optimized at these critical locations based on these simulation results.

A dynamic simulation of mechanical assemblies can be carried out as well as the static loading on the parts with finite-element models. The behavior of elastic materials in real time can be investigated. This typically requires long computation times. Simplified analogous models based on network analogies are often used for this purpose:

– Body masses are modeled as point masses placed at the center of gravity of the real object.
– Body elasticities are modeled as springs with a parameter for the spring rigidity.
– Damping characteristics of bodies are modeled as damper elements with a damping constant.

The decay response of the oscillation after an excitation can be analyzed for the exemplary rubber stop in Fig. 2.17 with the following dynamic analogous model

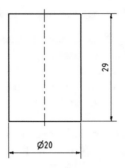

Fig. 2.20 Dimensioned drawing view of a part (rubber sleeve)

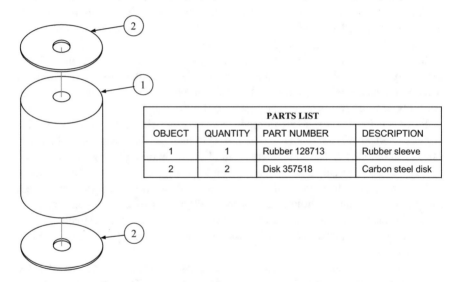

PARTS LIST			
OBJECT	QUANTITY	PART NUMBER	DESCRIPTION
1	1	Rubber 128713	Rubber sleeve
2	2	Disk 357518	Carbon steel disk

Fig. 2.21 Exploded view of an assembly (rubber stop) with parts list

(Fig. 2.23). The carbon steel disks are simplified as rigid elements and the rubber sleeves as massless elements. Some of the required concentrated parameters for the network elements, such as the masses of the individual objects, can be taken from the CAD model. Other parameters, such as the spring rigidity as a quotient of the pressure force and deformation, can be determined from the static finite-element simulations. The damping constant of rubber has to be obtained from the material data or the measurements.

The geometric and material properties of the CAD model are iteratively modified with the results of these simulations. The future characteristics of the entire

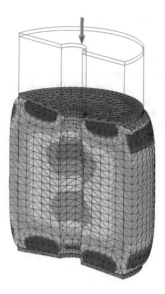

Fig. 2.22 A finite-element simulation where the stresses in the rubber sleeve are determined for compressive forces at the hole edges

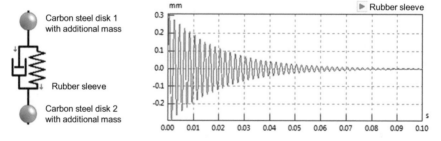

Fig. 2.23 Dynamic analogous model of the rubber stop in Fig. 2.17 with simulation results, showing the decay response of its oscillations after an excitation

electronic/mechanical system can thus be determined using the CAD models, which enables us to develop and simulate the desired optimal solution before undertaking its actual physical implementation.

References

1. G. Pahl, W. Beitz, J. Feldhusen, K.-H. Grote, *Engineering Design: A Systematic Approach*, Springer, 3rd edition. 2007
2. F. Kramer, *Innovative Produktpolitik, Strategie—Planung—Entwicklung—Einführung*, Springer, 1987

3. H. Kerzner, *Project Management: A Systems Approach to Planning, Scheduling, and Controlling*, Wiley, 11th edition, 2013

4. ISO 29845:2011, *Technical product documentation—Document types*, online at www.iso.org

5. ISO 128–30:2001, *Technical drawings—General principles of presentation—Part 30: Basic conventions for views*, online at www.iso.org

6. ISO 5455:1979, *Technical drawings—Scales*, online at www.iso.org

7. ISO 7200:2004, *Technical product documentation—Data fields in title blocks and document headers,* online at www.iso.org

8. ISO 128-40, 44, 50:2001, *Technical drawings—General principles of presentation—Part 40: Basic conventions for cuts and sections, Part 44: Sections on mechanical engineering drawings, Part 50: Basic conventions for representing areas on cuts and sections,* online at www.iso.org

Chapter 3
System Architecture and Protection Requirements

Having dealt with the major steps in the development process and the necessary drafting skills, we now move on in this chapter to the system itself, i.e., system-level functions and structures (Sect. 3.1), design variants (Sect. 3.2), and various technological implementations (i.e., electronic system levels) for a design solution (Sect. 3.3). These implementation options allow a development engineer to identify opportunities for modularization early in the design process to reduce costs and shorten the development timeline.

System protection issues should also be considered during development (Sect. 3.4). Each module and the electronic system must be designed so that it does not pose a risk to humans or to the environment. In addition, compliance with statutory requirements, such as, protection against electric shock (protection classes), protection against accidental contact, and protection against ingress of foreign objects or water (IP codes), is mandatory.

3.1 Introduction—Terminology, Functions and Structures

Electronic systems are functional and technical units whose operation allows the requirements of a technical system definition to be met. The primary characteristics of such systems include information flow, comprised of information acquisition, processing, transmission, storage, and output. There are also systems, consisting primarily of energy and material flows, in medicine, laboratories, and in the household, which are not typically classified under mechanical engineering (as "machines") due to their miniature size and, hence, are considered here.

Each system performs an overall function and typically is comprised of modules and components. *Modules* are self-contained complex units that function largely independently. On the other hand, *components* are parts that cannot be broken down further during development.

© Springer International Publishing AG 2017
J. Lienig and H. Bruemmer, *Fundamentals of Electronic Systems Design*,
DOI 10.1007/978-3-319-55840-0_3

3.1.1   System Characteristics of Devices

Complex devices, such as smartphones, and modules, such as printed circuit boards, are treated as generic systems to generalize the design process. By applying systems theory, devices and modules with different operating modes and complexities can be treated consistently and developed using common processes.

A technical system may be characterized by its relationships with the environment E, required functions F and a structure S.

- The system's *environment E* is the totality of objects and physical quantities outside the system that relate to the system. These environmental relationships are implemented as inputs and outputs.
- The *function F* is a system property designed for a given purpose that converts the necessary inputs into defined outputs under predefined environmental conditions. A printer's ability to produce a graphical image from binary signals is an example of such a property.
- The *structure S* of a system is the totality of its elements and their interrelationships.

Certain elements and relationships in the structure are activated by functional inputs and outputs, which execute the required function based on their characteristics. Hence, the technical function characterizes the relationship between the structure and the environment.

The development engineer must clearly define the characteristics E, F, and S to describe a product satisfactorily in the technical requirements document (functional specification, see Sect. 2.3.5).

3.1.2   System Environment

An electronic system must engage with its environment to function properly. Environmental relationships can be described by the general functional model (Fig. 3.1).

There are three interface categories between the system and environment that are keys for system functionality and configuration. The development engineer needs to envisage different environmental scenarios and implement the resulting functional and technical requirements accordingly.

Processing layer (interface 1)
The purpose of a system is to execute, effect, or communicate predefined technical operations. The system performs these operations internally by converting inputs I_P into outputs O_P. This activity is called the system *processing function*.

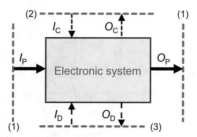

Fig. 3.1 Schematic view of the input (I) and output (O) relationships between a system and its external interfaces (processing quantities I_P, O_P; communication quantities I_C, O_C; and disturbances I_D, O_D). Interfaces (1) to (3) represent three interface categories related to the processing, communication, and security functions of the system (see Sect. 3.1.3 and Fig. 3.2)

Communication layer (interface 2)
An information exchange may occur between the system and humans or other technical products in technical systems. This operation takes place with communication inputs I_C for guiding or controlling the processing function, and with communication outputs O_C as feedbacks for monitoring this function. These operations are performed by a system's *communication function*.

Disturbance layer (interface 3)
All non-functional inputs that impact (typically negatively) the system as disturbances I_D, and the environment as outgoing disturbances O_D from the system, are handled in this layer. These disturbances are treated as independent variables. Suitable measures need to be put in place to counteract these quantities, to secure the system processing function, as well as sustain the required environment conditions. These operations are part of the system *security function*.

3.1.3 System Functions

The number of functions a system can perform corresponds to its number of useful physical characteristics. If the system fulfills many technical functions, the development engineer needs to consider the relationships between such functions. The processing, communication, and security functions mentioned above must be identified based on the interfaces between system and environment (Fig. 3.2).

– The *processing function* converts input functional quantities into output functional quantities under given environmental conditions. The conversion is executed via collaboration between hardware and software. Information, energy, and material flows all play a role in this function; as stated earlier, the information flow typically dominates in electronic systems.

Fig. 3.2 System functions
(*I* input, *O* output) with
processing quantities I_P, O_P,
communication quantities I_C,
O_C, and disturbances I_D, O_D.
Quantities D_I, D_O are internal
disturbances, *C* signifies
internal control quantities, and
M signifies internal
monitoring quantities [1]

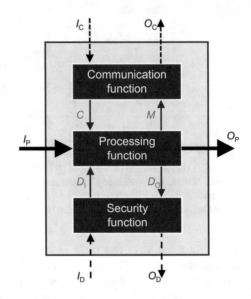

There are two types of processing functions: primary and secondary.
Information processing is generally the primary (main) processing function in
electronic systems, with a secondary energy and material processing function. If
we take a printer, for example, the primary processing function is the conversion
of electronic data into printed alphanumeric and graphical information (information processing), while blank paper is transformed into printed paper
(material processing) and electrical energy into mechanical energy (energy
processing) by secondary processing functions.

- The *communication function* is the information link between the device and the
  user or other technical systems to execute the processing function. The communication function also monitors the processing function by converting
  internal monitoring quantities into external supervisory and monitoring
  quantities.
- The *security function* performs three tasks, namely (1) protecting the processing
  function from possible environmental disturbances; (2) protecting it from
  system-internal disturbances; and (3) protecting the environment from system
  faults.

3.1.4 System Structure

The system structure, i.e., the device (system) configuration, is a prerequisite for
carrying out the system functionality. The structure consists of the system's parts,
the so-called *elements*, which are not further decomposed within the system, and
relationships, which describe the relationships between the system elements.

The development engineer should decompose the system into complexity levels depending on the specific design task. For example, when designing a system with complex functions, the engineer should treat complex purchase parts as single elements without breaking them down any further.

A structure comprises the following complexity levels:

- *Complex systems*, composed of various systems, e.g., a monitoring system;
- *Single systems*, comprising modules and components of varying degrees of complexities, such as a monitor;
- *Modules* as stand-alone groups of components that are linked together (modules can also be viewed as components if they are supplied fully assembled, e.g., integrated circuits or switches);
- *Components*: these are individual elements—such as resistors—that cannot be broken down any further.

3.2 System Design Architecture

An electronic system, as explained in Sect. 3.1, is composed of modules and components. Modules, such as printed circuit boards, are separate, stand-alone groups of interconnected components. A system can be designed in a variety of ways depending on the modules' functional and manufacturing requirements and the intended applications. The development engineer should define the design architecture early in a project, typically at the system-definition stage.

Design architectures are influenced by

- The granularity of the system configuration, i.e., the number of different elements in the system (Sect. 3.2.1),
- How the system is to be assembled (Sect. 3.2.2), and
- How the system is to be integrated in its environment (Sect. 3.2.3).

3.2.1 System Granularity

The following design architectures are available to suit the degree of system granularity:

- No pre-designed modules are used in the *custom-assembled* or *compact design approach*. In this case, the system is designed as a fully assembled unit using individual, custom-designed components. Small and compact dimensions can be achieved with this approach; however, the amount of work involved is much greater than with the following approaches.

Table 3.1 Standard levels and applicable norms for the 19-inch rack system in electronics

Level	Standard items	Standard
Level 1		
– Printed circuit boards	Printed circuit boards: printed circuits, substrates, grids, holes, nominal sizes, PCB measurements	IEC 60097, IEC 60249, IEC 60297
– Components	Components	IEC 60326
– Connectors	Connectors	IEC 60603
Level 2		
– Modules	Modules: PCB, cassette, plug-in package	IEC 60297
Level 3		
– Front panels	Front panels: width 482.6 mm (19″), subrack heights and mounting dimensions, rack installation dimensions	IEC 60297
– Subrack	Subrack: dimensions with indirect connectors	IEC 60297
Level 4		
– Enclosure	Enclosure: installation dimensions, enclosure stack	IEC 60297
– Racks	Racks: installation dimensions	IEC 60297
– Panels	Panels: panel dimensions, frame row pitch	IEC 60297

- The device is composed of self-contained functional modules in the *modular design approach*. This method provides significant design rationalization, for example, specific standard modules can be deployed with different device type series, and individual modules can be modified for rapid development as required.
- Extending the modular design approach across multiple companies by means of standardization allows *industry-wide standard module development*. Various different systems can be built based on a limited number of such module types. These standard modules are common, recurring system elements that are optimized and streamlined for combinability and reusability.

The *19-inch rack system in electronics* is a prime example of the latter standardized modular design approach that has seen widespread use in industry since 1934 [2]. Its key features, such as standard levels with underlying applicable norms, are listed in Table 3.1.

3.2.2   System Assembly

The design architecture can also be classified based on how the system is to be assembled (Fig. 3.3):

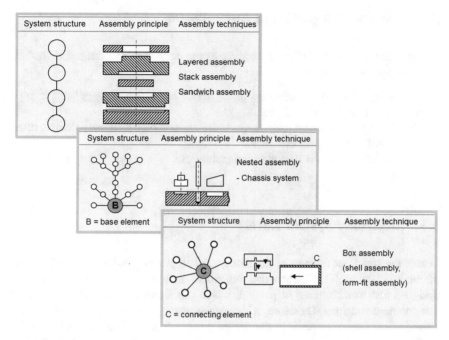

Fig. 3.3 Different assembly methods of an electronic system: the conventional layer, stack or sandwich assembly techniques (*top*), the nested assembly technique (*middle*), and the box assembly technique (*bottom*)

– Each component is put in place on the base or on the component that was fitted last in the *layered*, *stack* and *sandwich assembly techniques.*
– Components are largely put in place side by side with the *nested assembly technique* (also: *chassis system*). Components are usually placed in a frame or base element (in a chassis or a printed circuit board) that holds them. Components can be placed in any order in the rack with this system.
– With the *box assembly technique* (also: *shell* or *form-fit assembly*), components are held in place in the base and secured in position and orientation by a main connecting element (retaining element).

   This list of assembly methods is not exhaustive and not only determines how a system is assembled, but also how easy it is to dismantle (Sect. 7.5). It should be noted that, due to their hierarchical structure, many electronic systems use a combination of the above assembly methods.

3.2.3   System Integration in Environment

Systems can also be classified according to the way they are integrated in their surroundings, for example:

– The *panel-mounted* or *surface-mounted device* has fasteners that enable it to be fitted into a higher-level device.
– The *desktop* or *floor-standing device* is an autonomous, stationary device, often with stabilizing support elements.
– The *portable device* is designed for mobile use.

3.3   Electronic System Levels

One of the stand-out features of the modern development process is its high degree of modularization. This means that functional groups in a device are typically designed and manufactured in parallel, which can significantly reduce the overall development timeline. Therefore, a development engineer should identify modularization opportunities early in the design process to save time and costs.

Electronic functional groups are functionally and technically separate, stand-alone units, whose functional elements are primarily based on the effect of electrical quantities, such as current and voltage. These functional groups have made rapid progress in recent years as the development of their underlying technologies, such as at the circuit level, allows exponentially increasing functional densities (Moore's law).

System levels have emerged as a useful grouping in the design of a system or a module. They characterize the respective level of complexity of the associated electronic functional groups (Fig. 3.4).

Each system level contains different functional groups:

– *Discrete components* are self-contained units manufactured to perform an elementary electrical function. They include passive components, such as resistors, capacitors, and inductances, and active components such as transistors.
– *Integrated circuits* (IC) are composed of many elements permanently connected electrically and mechanically to form a functional unit. Integrated circuits may come as packaged (with enclosure) or unpackaged (bare dies).
– Bare dies and discrete components are connected electrically and mechanically to form a functional and technical unit in *multi-chip modules* (MCM). Historically, multi-chip modules are derived from hybrid modules, whereby unpackaged integrated circuits based on different technologies are combined with discrete components to create new functions, which are too expensive or cannot be built using (monolithic) integrated circuit technology.

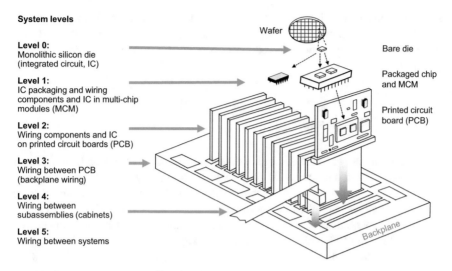

Fig. 3.4 Different system levels in an electronic system [3]

– Different components, including integrated circuits and multi-chip modules, are connected together electrically on *printed circuit boards* (PCB). Printed circuit boards are typically comprised of a base material as substrate, with added copper layers to connect components and integrated circuits electrically. In addition to rigid printed circuit boards, flexible printed circuit boards are increasingly being deployed.
– *Power-supply elements* are functional units for providing a custom energy supply, typically involving energy conversion, such as voltage transformation or rectification. There are three types of power-supply units: (1) grid-powered, (2) non-interruptible, and (3) autonomous.
– The purpose *of external electrical interconnects* is to connect modules and systems to enable energy and information flow between them.

3.4   System Protection

Interactions between the system (device) under development and the environment should be considered early in the design process. Device reliability and protection criteria are more challenging today due to the broad application of modern systems, coupled with the potential for extreme climatic conditions that may be encountered. The requirements for successful product development include protection against electric shock, protection against the ingress of foreign bodies and water, protection against high temperatures (climate protection) and radiation (electromagnetic compatibility), protection against the effects of implosion, inadequate stability, and injuries caused by moving parts, and protection against fire. Specialist knowledge is required for protection classes, IP codes, and thermal stress (Chap. 5) and also increasingly for protection in connection with electromagnetic compatibility (Chap. 6).

3.4.1 CE Designation

A product may only be sold and put in operation within the European Union (EU) if it complies with all applicable EU Directives. A conformity assessment as per these directives must be performed for electronic systems. Compliance is indicated by the vendor or importer with the *CE marking* (Fig. 3.5).[1] The CE marking is also found on products sold outside the European Union that are manufactured in (or designed to be sold in) the EU. This makes the CE marking recognizable worldwide.

Every vendor or distributor of an electronic device within the EU, who puts the device into service, declares, by attaching the CE marking in a *Declaration of Conformance* (EU declaration of conformity), that the device complies with all EU Directives. This requires that the engineers who developed the device have awareness of the respective EU laws. Essentially, the development engineer is responsible for the correctness of this stamp of approval and thus, for compliance with all relevant safety regulations and guidelines. An external test is only required for devices that have increased risk potential.

The CE marking certifies that a product may be brought into service only when it does not pose a risk to the health and safety of users or third parties. It is important to note that this safety requirement also applies if the device is used incorrectly in a foreseeable manner.

3.4.2 Protection Classes

One of the key aims of equipment protection is to protect the user against electric shock; this is achieved by the use of *protection classes* (also called *appliance classes*) [4]. They specify measures to prevent hazardous contact voltages on *unenergized* parts of an electronic device. All electronic systems (devices) must comply with one of the following three protection classes:

- *Protection class I* is not only based on the basic insulation, but also conductive parts are also connected with a low-resistant protective conductor, also called protective grounding. If, in the event of a fault, a current-carrying wire makes

---

[1]It is said that "CE" originated as an abbreviation of Conformité Européenne, meaning European Conformity, but this claim has never been officially confirmed. Other sources refer to "CE" as Communauté Européenne (European Union).

Fig. 3.5 CE marking is the manufacturer's declaration that the product meets the requirements of the applicable EU directives. The mark consists of the CE logo and, if applicable, the four digit identification number of the notified body involved in the conformity assessment procedure

Protection class I Protection class II Protection class III

Fig. 3.6 Symbols for designating protection classes as defined in the international standard IEC 61140 [4]

contact with the protective conductor connected to the enclosure, a high current is able to flow in the "grounded" protective conductor, which triggers the fusible cut-out, which in turn de-energizes the device. These types of devices are fitted with a device connector with grounding contact, where the low-impedance of the grounding protective conductor between enclosure and protective conductor at the terminal plug must meet certain standards and requirements. The connection to the protective conductor should be the first connection that is made when the plug is plugged in, and it should be the last to be broken when the plug is removed.
– *Protection class II* not only provides basic insulation, but also reinforced protective insulation. High insulation resistances are required for predefined test voltages. The devices should not be connected with the protective conductor of the installation, as protection is unlikely to be provided when the fusible cut-out is triggered due to low-fault currents caused by the high-impedance enclosure.
– *Protection class III* offers protection by the exclusive use of a safety extra-low voltage (SELV), which is (in many countries) 50 V for AC (rms value), 120 V for DC, 24 V for children's toys, and 6 V for medical systems for application in the body. Devices with safety extra-low voltage are operated without a protective conductor and should not be connected to the protective grounding of the extra-low voltage generator or to other (live) parts under voltage.

The symbols depicted in Fig. 3.6 are used to designate systems with the respective protection class.

Appliances with protection class 0 are prohibited in much of the world for safety reasons as they have no protective-earth connection and feature only a single level of insulation.

Table 3.2 Protection details designed for IP codes comprising at least two digits x_1 and x_2. They are defined in the international standard IEC 60529 [5]

Digit	First digit x_1		Second digit x_2
	Shock protection	Protection against ingress of foreign bodies	Protection against ingress of water
0	No protection	No protection	No protection
1	Protection against random contact with the hand; protection against access to dangerous parts with the back of the hand	Protection against solid foreign bodies 50 mm and more in diameter	Protection against vertically dripping water
2	Protection against contact with the fingers; protection against access to dangerous parts with the fingers	Protection against solid foreign bodies 12.5 mm and more in diameter	Protection against vertically dripping water when the enclosure is tilting at an angle up to 15°
3	Protection against contact with tools; protection against access to dangerous parts with a tool	Protection against solid foreign bodies 2.5 mm and more in diameter	Protection against spraying water
4	Protection against contact with tools and wires; protection against access to dangerous parts with a wire	Protection against solid foreign bodies 1.0 mm and more in diameter	Protection against splashwater from any direction
5	Complete protection against contact; protection against access to dangerous parts with a wire	Protection against dust (dust-proof)	Protection against water jets
6	Complete protection against contact; protection against access to dangerous parts with a wire	Protection against dust (dust-proof)	Protection against high-pressure water jets
7	–	–	Protection against temporary immersion in water
8	–	–	Protection against continuous immersion in water

3.4.3   IP Codes of Enclosures

The system protection also includes the protection of the device interior by the mechanical casings and electrical enclosures of the system. This protection is called *IP Code*, *International Protection Marking*, sometimes interpreted as *Ingress Protection Marking*. The IP Code is typically defined as IP x_1 x_2 and subdivided into protection provided against intrusion (body parts such as hands and fingers), protecting against ingress of foreign bodies, and protection against ingress of liquids [5].

Table 3.3 Examples of IP codes

Application	Operating conditions	Typical locations	Protection type
Low-level protection	Dry building interiors with no condensation	Offices, apartments, and business premises, showrooms	IP 10, IP 20
Average protection	Building interiors where condensation can occur or devices in farm vehicles	Kitchens, deep freeze rooms, cellars, closed stables, automobiles, closed trucks	IP 30, IP 40, IP 41, IP 22
Medium protection	Outdoor rooms, temporary outdoor operation	All weather shelters, tents, roofed surfaces, open-cast mining	IP 22C, IP 34, IP 43, IP 44
Strong protection	Continuous operation outdoors	Locations where the harmful effects of ambient weather conditions are permanent	IP 54, IP 56, IP 65, IP 66
Total protection	Temporary or permanent flooding with water		IP 67, IP 68

The protection details specified with at least two digits x_1 and x_2 for the IP protection types are shown in Table 3.2. The first digit indicates on a scale of 0–6, the level of protection that the enclosure provides against access to hazardous parts (e.g., electrical conductors, moving parts). It comprises the protection against intrusion of body parts such as fingers and hands (shock protection) and the ingress of solid foreign bodies such as dust. The second digit indicates on a scale of 0–8, the level of protection that the enclosure provides against harmful ingress of water. Table 3.3 lists examples of these classifications.

The capital letter "X" is used instead of a digit when no particular protection is specified (note that it should not be confused with "no protection"). IP protection type of at least IP $3X$ is mandatory for a device with live parts over 50 V AC or 120 V DC in most countries.

Two further positions, in addition to the double-digit, numerical IP protection types, are available for extra information to form a four-position IP $x_1x_2x_3x_4$. The third position can contain the letters A, B, C, or D to define extra protection against contact with dangerous parts with the back of the hand (A), the finger (B), a tool (C) or a wire (D). An additional letter at the fourth position provides further information for high voltage devices (H), whether the moving parts were in motion (M) or at a standstill (S) during the water test or whether the test was carried out under predefined weather conditions (W).

References

1. W. Krause, *Gerätekonstruktion in Feinwerktechnik und Elektronik*, Hanser Verlag, 2000
2. G. R. Mezger, "The Relay Rack in Amateur Construction", pp. 27–30, *QST*, vol. 18 (1934), American Radio Relay League, Hartford, CT, 1934

3. R. R. Tummala, *Fundamentals of Microsystems Packaging*, McGraw-Hill, 2001
4. IEC 61140:2016, *Protection against electric shock - Common aspects for installation and equipment*, Publication date 2016-01-07
5. IEC 60529:1989+AMD1:1999+AMD2:2013 CSV, *Degrees of protection provided by enclosures (IP Code)*, Publication date 2013-08-29

Chapter 4
Reliability Analysis

Inexperienced engineers typically neglect reliability issues when designing. Reliability parameters are sometimes not known, and as a result the development engineer often maximizes them, "to be on the safe side." However, the resulting excessive costs often require a redesign. On the other hand, when reliability data for a system are provided, they raise questions on the required reliability of the system's individual components.

This chapter explains the mathematics needed to perform reliability analysis and introduces the primary reliability parameters, which a development engineer must be familiar with today (Sects. 4.1 and 4.2). The so-called bathtub curve of failure rates is applied for the reliability of electronic systems; understanding the middle of this curve, which represents a constant failure rate, is critical in practice. The reliability parameters for electronic systems are easily calculated by applying this constant failure rate, which is associated with an "exponential failure distribution" (Sect. 4.3). The failure modes of electronic components are described in Sect. 4.4. We show how the required reliability parameters for individual components and modules can be determined by applying the exponential failure distribution to these failure modes (Sects. 4.5 and 4.6). In addition, we show how the system reliability can be calculated from the reliability of individual components.

Finally, Sect. 4.7 contains recommendations for upgrading the reliability of electronic systems.

4.1 Introduction

Function and reliability are the two most important factors impacting system quality. An electronic system should fulfill its required *functions* based on given parameters (output/response values) within defined boundaries. *Reliability* is a measure of the performance of these functions over a given period. The parameter

© Springer International Publishing AG 2017
J. Lienig and H. Bruemmer, *Fundamentals of Electronic Systems Design*,
DOI 10.1007/978-3-319-55840-0_4

boundaries are determined by the predefined operating modes and conditions of usage, along with maintenance, storage, and transportation.

High reliability is especially important for electronic systems in industrial equipment, where a device failure often results in production loss or rejects. The associated costs in such cases may be higher than the initial outlay for the defective system.

We will show later how the product *price* increases as the reliability increases. The price increase associated with increased reliability is, however, offset by a reduction in extra costs for the customer for maintenance and repair during the product lifetime. Thus, customer costs can be *optimized* by examining the relationship between the overall costs and system reliability (Fig. 4.1).

If there is a loss of reliability, *maintenance costs* may exceed customer expectations or the proposed manufacturer guarantee and warranty costs. In addition, the business's reputation may also suffer and it may lose customers if the reliability of its products is lower than that of its competitors' products. On the other hand, if a product's reliability exceeds market expectations, without offering additional technical benefits, it can become so expensive that it is not profitable to manufacture.

Against this backdrop of cost efficiency, a product need not be overly reliable. Rather, the development engineer should aim to minimize overall costs by seeking reliability that fits the purpose. There are exceptions to this rule, however. Maximum reliability is required in aerospace engineering, for example, as maintenance and repair are extremely difficult, if not impossible. The same applies to healthcare systems: health and safety risks and environmental concerns prohibit cost cutting.

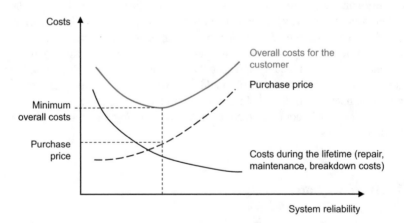

Fig. 4.1 Achieving cost efficiency by examining the relationship between the overall costs and system reliability. The increased reliability associated with higher purchase prices is offset by a reduction in repair and maintenance costs, resulting in an overall cost optimum at specific (to-be-aimed-for) reliability values

Reliability data are always future oriented. The proper functioning of a system can only be predicted with a specific probability, as the reliability parameters are stochastic (random). Nevertheless, every development engineer needs to define the target reliability level for the system at hand.

The functionality, accuracy, processing power, etc. of electronic systems are constantly improving. At the same time, due to these increased levels of complexity, there is also an increased risk of systems becoming prone to failure. These factors cause a conflict of interests. A culture of reliability is needed to accompany systems throughout their entire life cycles. Reliability must be designed "into the systems," starting with product planning, and on through component selection and the system design itself, to fabrication and quality control. Since reliability should be "designed in," it should be considered a strategic task. (In contrast, maintenance, keeping components and systems functioning, is considered a tactical task.)

The engineer remains responsible for reliability when the product moves from engineering into production. The developed system is also tested "in the field," as the product is still under warranty and the manufacturer guarantees the quality of its product for the customer. Failures, which are not random failures, must be carefully assessed for a period of several years. The development engineer remains responsible for the reliability parameters and calculations even years after the design and development is complete.

4.2 Calculation Principles

4.2.1 Probability Terminology

Probability plays a crucial role in reliability theory. We will introduce probability here with the game of dice, using a single cubic die whose six faces represent the numbers 1–6.

Events can be certain, impossible or random. An event E is *certain* if it always occurs under given conditions. It is an *impossible* event if it can never arise. A *random* event can occur or not occur.

Let us say the random event E "4 on top" occurs 15 times for a 100 rolls of a die. The *relative frequency* $H(E)$ of the events E in the series of rolls of the die can be expressed as follows:

$$H(E) = \frac{m}{n}, \tag{4.1}$$

where the parameter m is the number of occurrences of E for n tries. This yields the fraction $15/100 = 0.15$ in the above example.

A regular pattern emerges when the relative frequency of an event E is calculated for a large number of tries. The relative frequencies for the numbers 1–6 on a die

Fig. 4.2 Relative frequency
$H(E)$ of individual numbers
1–6 for different numbers of
die rolls

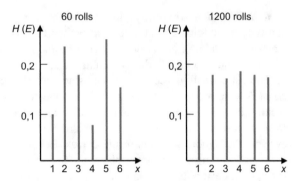

approach a fixed value, namely 1/6 for an "ideal die." Deviations from this value decrease as the number of throws increases (Fig. 4.2).

This value is called the *probability* of event E, i.e., the probability of rolling a particular die number. In general, the probability of an event is the relationship of the given (favorable) results to the total set of possible results. Its value range lies between 0 (event never occurs) and 1 (event always occurs), where percentage values between 0 and 100% are often used as well.

Probability theory applies when the same object is repeatedly observed under the same test conditions. For a sufficiently large number of tests n, where the event E occurred m times, the relative frequency m/n can be used to estimate the probability $P(E)$.

Not only can probabilities be determined experimentally (subsequently), they can be theoretically predicted as well (in advance). The *probability of occurrence E* of an event is a predetermined theoretical value. Actually meeting the probability of occurrence E is a function of the number of tries; the higher the number of tries, the more likely the observed, relative occurrence will fit the predicted one (see Fig. 4.2).

There is a difference between the probability of a single event and the total probability, the latter of which may increase or decrease over multiple events, depending on the nature of the event. For example, the probability of occurrence that the number 4 turns up when the die is rolled a single time is 1/6, 0.1667 or 16.67%.

Obviously, the probability of rolling a given number increases as the number of tests increases (such as by throwing a number of dice simultaneously, or throwing a single die a number of times in succession). This total probability is determined from the product of many single probabilities; for example, the product $1/6 \cdot 1/6 = 1/36$ represents the probability that the number 4 is thrown *twice*. However, the scenarios of interest here, i.e., *rolling the number 4 (at least) once* cannot be determined with this approach. The focus instead is on the complementary event (not rolling number 4) and its probability. This is 5/6 for one try, and $5/6 \cdot 5/6 = 25/36$ for two tries. Since this is the probability of not rolling a 4, we must subtract this value of 25/36 from 1 to arrive at 11/36 as the probability of throwing the number 4 (at least) once in two tries.

The same procedure is followed for *n* tries by multiplying the probabilities of the complementary event *n* times and subtracting the result from 1. Later, we will see how this technique is applied to determining the failure probability of systems, which is based on the individual failure probabilities of their functional components. The single probabilities of the complementary event (component does not fail) are multiplied together here as well to calculate the probability of survival of the system. The failure probability of the system is obtained by subtracting this survival probability from 1 (Sect. 4.6.3).

4.2.2 Reliability Terminology

Reliability is defined as the property with which a product performs

– a required function,
– under defined functional and environmental conditions,
– during a given service period.

Reliability is therefore a *probability* between limit values 0 and 1, in which the operating period plays a fundamental role. Please note that every system or device will fail during a sufficiently long operating period, and thus the reliability approaches 0. Reliability data are meaningless if they are not based on a given operating period. Furthermore, the system must be operational at the beginning of the reliability analysis.

The serviceability of components and systems depends greatly on their operating environment; on compliance with preset maximum stress levels and loading; and on environmental conditions, such as temperature, humidity, and vibrations. Thus, functional and environmental conditions for the product typically need to be specified as well as reliability data. The development engineer selects and specifies these conditions.

A module or system is said to have *failed* if its intended purpose is not met, i.e., at least one parameter (output value) exceeds the minimum or maximum limit value. In the case of failure, the system function must be reinstated by an unplanned external action, such as a *repair*. When an external action is planned, for example, to replace consumables, it is called *maintenance*. Maintenance, in contrast to repair, is a preventative measure, typically to counteract wear and tear.

4.2.3 Reliability Parameters

The reliability of components and systems is defined as the probability of proper operation under the required operating conditions and for a defined period of time. It is expressed by parameters that describe their response over time with respect to failures, and their prevention and repair. These parameters are mean values of a totality, probability data, or probability-related prognoses (probability of occurrence). Reliability is typically reduced to a quantitative reliability metric, such as failure rate, which is introduced below.

The key parameters will be illustrated using the life expectancy graphs for humans shown in Fig. 4.3. These graphs show various functions that approximate the relationship between a reliability parameter and time.

The *reliability function* (also: *probability of survival* or *survival function*) $R(t)$ captures the probability that a system will survive up to a specified time. It is generally defined as follows:

$$R(t) = \frac{n(t)}{n_0},\tag{4.2}$$

where $n(t)$ represents the failure-free units to time t, and the initial stock n_0, which represents the totality of working units at the starting time t_0.

Only 50% of the initial stock of the human group shown in Fig. 4.3 are alive after 76 years. In other words, a given human being has a 50% chance of living to be 76 years old. Only the probability of survival, which drops over time, can be determined. We cannot predict when a given person will die.

The arithmetic mean value of the reliability function $R(t)$ is calculated from the areas above and below the curve $R(t)$ (Fig. 4.3, top). This mean value m for repairable units is called the *mean time between failures* (*MTBF*), and the *mean time to failure* (*MTTF*) for non-repairable units is:

$$m = \int_{0}^{\infty} R(t) \cdot dt.\tag{4.3}$$

The *MTBF* (*MTTF*) value can be used as a system reliability parameter or to compare different systems or designs. This value, as we will show later, should only be understood conditionally as the "mean lifetime" (an average value), and not as a quantitative identity between working and failed units. The mean time to failure is $m = 73$ years for the given human group shown in Fig. 4.3, with approximately 55% of the initial stock still alive.

The *failure distribution function* (also: *failure probability* or *unreliability function*) $F(t)$ is the probability that an element will fail before time t. It is thus the complementary quantity of the reliability function $R(t)$ and is expressed as follows:

$$F(t) = 1 - R(t).\tag{4.4}$$

Due to this relationship, it is sufficient to know one of the two functions as a function of time. For example, one can observe the fraction of a population of components that fail at time t in order to estimate the value of the failure distribution function at this time.

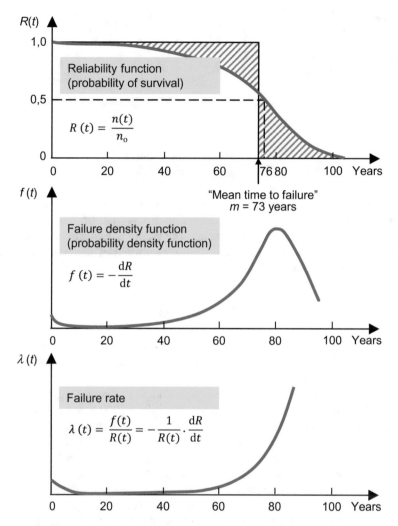

Fig. 4.3 Reliability-related functions for a human group to illustrate the terms *reliability function*, *failure density function*, and *failure rate*

The *failure density function* (also: *probability density function, PDF*) $f(t)$ is the mathematical derivative of the failure probability and thus defines how the failure distribution function $F(t)$ changes with time:

$$f(t) = \frac{\mathrm{d}F(t)}{\mathrm{d}t} = -\frac{\mathrm{d}R(t)}{\mathrm{d}t}. \tag{4.5}$$

The *failure rate* $\lambda(t)$ is the probability that a module that has not failed up to time t will fail up to time $t + \mathrm{d}t$ (i.e., in the small period $\mathrm{d}t$):

Fig. 4.4 Typical plot of the failure rate $\lambda(t)$ over time t (bathtub curve) with decreasing, constant, and increasing failure rate sections

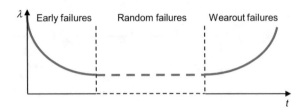

$$\lambda(t) = \frac{f(t)}{R(t)} = -\frac{1}{R(t)} \cdot \frac{dR(t)}{dt}. \qquad (4.6)$$

The failure rate thus defines how many units will fail on average in a time unit. This critical reliability parameter is defined in failures per time unit, that is, the reciprocal of time, e.g., h^{-1} (failures/hour). The failure rate cannot be measured for an individual unit, and it is estimated from observations of the failure mode of a larger number of similar units:

$$\lambda(t) \approx \frac{\Delta n_a}{n(t) \cdot \Delta t}, \qquad (4.7)$$

where Δn_a is the number of failures in the given interval Δt, $n(t)$ the number of all units functioning at the beginning of the analysis, i.e., at time t, and Δt is the observed time interval.[1]

We will illustrate this estimate with two examples. If an average of five components from a set of 100 working components fails in 1000 h, the failure rate for a component is $5 \cdot 10^{-5}$ failures/hour. Should a stamping device temporarily fail 100 times in 1000 h uptime, the failure rate is 0.1 failures/hour (0.1 h^{-1}).

A plot of the failure rate over time is given in Fig. 4.4. This curve is known as the *bathtub curve*. This typical characteristic curve can be broken down into three sections *early failures, random failures,* and *late* or *wearout failures.*

Early failures, also known as *infant mortality fails* or *early life failure rates (ELFR)*, occur during the early life of a product due to manufacturing defects that escape detection. They are caused by inadequate quality management during fabrication and inadequate product testing. Typically, the failure rate drops during this period. The remaining products, presumably without manufacturing defects, should remain functional for their design lifetime.

Random failures occur during this expected product lifetime. They are unpredictable and unforeseeable and are due to the statistical superimposition of a

---

[1]Calculating the failure rate for ever smaller intervals of time, i.e., the *instantaneous* failure rate as Δt tends to zero, results in the *hazard function* (also called *hazard rate*), $h(t)$. As it is a function that describes the conditional probability of failure, it is always a value between 0 and 1. (Failure rate, as the count of failures per unit time, can be a value greater than one.)

number of independent factors. The failure rate is constant during this period of intrinsic fails, also known as the *intrinsic failure period* or the *stable failure period*.

Wearout failures, also known as *late failures*, occur at the end of the service life through wear and tear, fatigue, aging, etc. This period is characterized by an increase in the failure rate.

4.3   Exponential Distribution

4.3.1   Reliability Distributions

Reliability distributions of components and systems need to be modeled with suitable mathematical functions for them to be mathematically treated in practice. Standard statistical distributions are matched with empirically determined distributions of reliability parameters. The exponential distribution, the Gaussian normal distribution, or the Weibull distribution are such functions (Fig. 4.5).

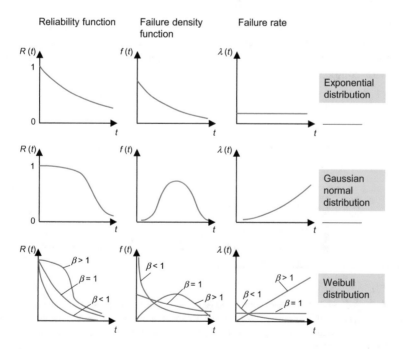

Fig. 4.5   Different distributions of reliability parameters over time. The decline in numbers due to random failures (i.e., constant failure rate) can be expressed using the exponential distribution. The Gaussian normal distribution is suitable to describe wearout failures (with electric motors, relays etc., and often with mechanical elements). Early and wearout failures can be modeled by the Weibull distribution with the shape parameter β

Complex reliability distributions can be approximately described as the summation of *Weibull distributions* that can be suitably adapted with their shape parameters to predefined characteristic curves (also: shape factor, Weibull slope). The Weibull distribution can be used to describe decreasing, constant, and increasing failure rates in technical systems. The exponential distribution with its constant failure rate described below is a special case. The "bathtub curve" (see Fig. 4.4) can be approximated by a summation of three Weibull functions. Due to its versatility, the Weibull distribution is one of the most widely used lifetime distributions in reliability engineering.

The *exponential distribution* is frequently used to model electronic components that usually do not wear out until long after the expected life of the system in which they are installed. It describes the decline in numbers caused exclusively by random failures occurring during the useful life of tested, *nonaging* electronic components. It covers the middle of the bathtub curve (Fig. 4.6). So-called memorylessness plays a key role here. The probability that a component that has been used for a number of days will last another *x* days is the same as that of a new component lasting *x* days. The exponential distribution is therefore not applicable to humans, and other living things, as the probability that a newly born child will live for 50 years is not the same as the probability that a 50-year old will live to be a hundred.

Of all reliability distributions, the exponential distribution is the most commonly used in electronic systems design, since such systems typically do not experience wearout type failures. It is also sufficiently accurate for reliability calculations of electronic components and systems. It is thus used exclusively throughout this book.

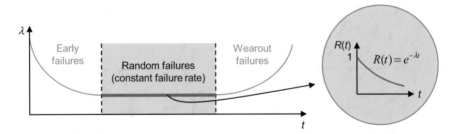

Fig. 4.6 Exponential distribution with its constant, low failure rate is restricted to the middle of bathtub curve; hence, it does not include early and wearout failures. The probability of survival $R(t)$ is indicated in this period by an exponential function, the exponential reduction in the number of operational units over time

4.3.2   Reliability Parameters and the Exponential Distribution

As noted earlier, the decline in population size can be described with an exponential function using a constant failure rate λ, i.e., purely by random failures. The reliability function $R(t)$ can be expressed as follows:

$$R(t) = \frac{n(t)}{n_0} = e^{-\lambda t} \tag{4.8}$$

for the special case $\lambda(t)$ = constant based on Eq. (4.6).

As introduced in Eq. (4.2), the number of failure-free units at time t is $n(t)$ and n_0 is the initial stock units. The failure distribution function $F(t)$ can thus be expressed as follows:

$$F(t) = 1 - e^{-\lambda t}. \tag{4.9}$$

The functions $R(t)$ and $F(t)$ are plotted in Fig. 4.7. The exponential function as per Eq. (4.8) can be described with the help of the parameter m. This value, as with the time constant of RC circuits, can be determined graphically via the initial tangent of the e function.

The other parameters are expressed as follows:

$$\text{failure (probability) density function} \quad f(t) = \lambda \cdot e^{-\lambda t}, \tag{4.10}$$

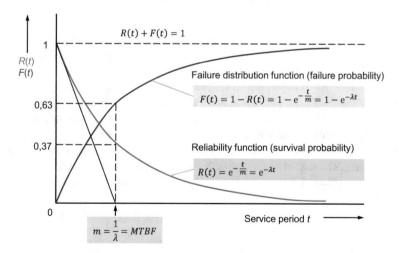

Fig. 4.7 Failure and reliability functions with the exponential distribution, i.e., during a period of constant failure rate. Also shown is the proportion of failed and operating units after a service period corresponding to the *MTBF* or *MTTF*

$$\text{failure rate} \quad \lambda(t) = \frac{f(t)}{R(t)} = \lambda = \text{constant}, \tag{4.11}$$

$$\text{mean time between failures } (MTBF \text{ or } MTTF) \quad m = \int_0^\infty e^{-\lambda t} \cdot dt = \frac{1}{\lambda}. \tag{4.12}$$

The mean time between failures (*MTBF*) and the mean time to failure (*MTTF*) are the mean value of the exponential failure distribution and, hence, are the reciprocal of the failure rate λ, according to Eq. (4.12). The reciprocal is defined mathematically as the area under the reliability function $R(t)$ curve. The units used are typically hours or life cycles. This critical relationship between time between failures (*MTBF*, *MTTF*) and failure rate allows a simple conversion when one of the two quantities is known and an exponential distribution (constant failure rate) can be assumed.

According to Eq. (4.8), only $e^{-1} \approx 0.37$, i.e., approximately 37% of the test units available at the beginning of the test (initial stock) are properly functioning after $t = MTBF = 1/\lambda$ hours have elapsed. In other words, the probability of an individual unit in a group surviving, during a given service period of *MTBF* hours, is only 37%. The probability of its failure up to this time is 63%. It is important to remember these relationships when using *MTBF* or *MTTF* for reliability predictions.[2]

4.4   Failure of Electronic Components

We have covered the typical relationship of the failure rate with time in Sect. 4.2.3 ("bathtub curve" in Fig. 4.4). Failure rates for electronic components drop to a few percent of their original values after a "burn-in period" of 300–500 h (2–3 weeks continuous operation) according to different sources. The end of the early failures should coincide with the product leaving the manufacturer's test area.

The exponential distribution with its simple mathematical formulas, introduced above, applies only to random failures, that is, to the linear part of the "bathtub curve." The phrase "random failures" is not very accurate, as every failure is caused by something. The notion of randomness should be viewed statistically and is based on the assumption that the time of failure for a given component cannot be predicted.

---

[2]Since *MTTF* can be expressed as "average life (expectancy)", many engineers assume that 50% of items will have failed by time $t = MTTF$. This inaccuracy can lead to bad design decisions. Furthermore, the above statements imply the total absence of systematic failures (i.e., a constant failure rate with only intrinsic, random failures), which is not easy to verify.

The failure rate for every component, and thus every system, becomes time-dependent sooner or later. The reliability then depends on the accumulated service period and cannot be described with the exponential function. Internal interconnects in electronic components often suffer fatigue due to reverse thermal stress and become increasingly prone to failure. In addition, the surrounding atmosphere or impurities may enter through defects in enclosures to alter the state of a component.

4.4.1   Drift

The definition of reliability includes requirements on the behavior of properties during a given period. When is a component said to have failed, for instance? While the requirements for digital components are fairly straightforward, analog components have drift issues. *Drift* can be defined as the gradual change in attributes over time that cause failure for a given loading. Drift is caused by chemical and physical processes inside a material or on its surface. The totality of these processes and their impact is called *aging* (Sect. 4.6.5; Fig. 4.12).

Drift failures can be differentiated between *monotonic* and *non-monotonic drift*. Monotonic drift is characterized by a parameter (output value) continuously varying in the same direction, crossing a predefined minimum or maximum value at some point. Non-monotonic drift is characterized by both positive and negative variance of a parameter.

Drift does not necessarily result in the failure of a functional unit, such as an interruption or short circuit. In the case of non-monotonic drift, it may be that the parameter quickly drifts back into its "acceptable region." Hence, the consequence of drift should instead be defined in terms of the resulting accumulated degradation. Such information, unfortunately, is often not sufficiently detailed in component data sheets. Many engineers do not appreciate the significance of these issues.

4.4.2   Reference and Operating Conditions

Functional and environmental conditions can also significantly impact component failure. Therefore, failure rates without reference to functional and environmental conditions have little or no value. Operating conditions arise from functional loads, such as current, voltage, and power dissipation; and environmental conditions include all component stresses originating in the environment. These include ambient temperature, pressure, humidity, radiation, and mechanical stresses, such as vibration and shock.

Failure rates quoted by component manufacturers are based on reference conditions, if not otherwise specified, using the concept of *base failure rates*. Examples of climatic and mechanical reference stresses relevant for these base failure rates are listed in Table 4.1.

Type of stress		Reference stress
Ambient temperature		40 °C (104 °F)
Relative humidity		30%
Air pressure for electronic components negligible in the range $(0.860–1.060) \cdot 10^5$ Pa		$1.013 \cdot 10^5$ Pa
Mechanical stress	Vibration stress	Frequency: (10–55) Hz
		Acceleration: 20 m/s^2
	Shock stress	Acceleration: 150 m/s^2
		Duration: 11 ms
Other stresses: wind, rain, snow, ice, dripping, spraying, splashing water or water jets, dust, the effects of pests, corrosive gases, radioactive radiation, temperature change, etc.		None

Table 4.1 Climatic and mechanical reference conditions for electronic components [1]

Influence factors (also: *reduction, loading,* or *stress factors*) are used to determine the failure rate under given functional and environmental conditions. They are based upon the intended operating environment of the component or system. These factors are defined as follows:

$$\lambda_B = \lambda_{ref} \cdot \pi_V \cdot \pi_I \cdot \pi_T \cdot \pi_E \qquad (4.13)$$

λ_B failure rate under operating conditions (operating failure rate)
λ_{ref} failure rate under reference conditions (base failure rate)
π_V voltage factor for voltage influence
π_I current factor for current influence
π_T temperature factor for temperature dependence
π_E environmental factor for environmental influences (factors other than temperature, such as humidity and vibration.).

There are other factors as well that are used as multipliers for the base failure rate. Some of these factors will result in a linear degradation of the base failure rate, while others, especially temperature-related factors, result in an exponential degradation.

4.4.3 Failure Rates of Electronic Components

The failure rate λ of electronic components can be approximated, using a method similar to that used in Eq. (4.7), as follows:

$$\lambda \approx \frac{\text{number of failures}}{\text{number of test units} \cdot \text{period}}. \qquad (4.14)$$

Influence factors π_n for determining the failure rate under operating conditions (λ_B) should be used when the failure rate λ was calculated under reference conditions (λ_{ref}) (Sect. 4.4.2).

The failure rate of electronic components is often defined by *Failure in time* (FIT). The FIT metric is the number of failures that occur in 10^9 h, which is approximately 114,000 years. The need for the unit "10^{-9} h^{-1}" illustrates the high reliability of today's electronic components, especially in microelectronics.

Table 4.2 shows reference values of failure rates for electronic components under reference conditions (base failure rates). These values are examples only, and should not be used in reliability calculations.

Table 4.2 Failure rates for electronic components under reference conditions [2]

Element	λ_{min} in FIT = 10^{-9} h^{-1}	λ_{max} in FIT = 10^{-9} h^{-1}
Transistors		
Thyristor, Triac	2.2	
Transistor, bipolar	0.74	
Transistor, FET	4.5	12
Diodes		
Signal	1	3.8
Z	2	
Power	5	
IC		
Digital, bipolar	2.5	80
Digital, MOS	10	290
Digital, CMOS	160	240
Resistors		
Carbon film	0.07	4.8
Wire	3.3	33
Thermistor	21	105
Potentiometer	8.9	650
Capacitors		
Electrolyte	9.5	2000
Film	0.53	490
Ceramic	0.67	23
Tantalum	2.1	240
Piezoelectric resonators	11	38
Interconnects		
Vias	0.041	0.26
Solder joint, automatic	0.069	
Solder joint, manual	0.14	2.6
Plug connection	0.12	
Crimped connectors	0.26	

Derating curves[3] are available for a variety of components, which depict the relationship between the base failure rate and the stress. Figure 4.8 shows typical curves for carbon composition resistors and dry tantalum capacitors. The rise in the failure rate as a function of temperature and electrical stress (power, voltage) is universally applicable to all electronic components. Influence factors (π) can be derived from these relationships.

The MIL-HDBK 217 [2] handbook is widely used to determine failure rates in the defense industry, aeronautics, and shipbuilding. The latest technological advances are not covered in the handbook as it has not been revised and updated since last issued in 1991 (revision F).

The Siemens standard SN 29500 [5] is a very popular resource for failure rates and is regularly revised and updated. The standard allows FIT values to be calculated based on known stress data for the most commonly used electronic and electromechanical components. This industrial standard is widely used for making reliability prognoses although it is not officially recognized globally. The failure rates acquired with the standard are conservative, and the resulting reliability errs on the safe side.

Exemplary failure rates are also cited in the application standard ISO 13849-1 [6].

4.4.4   Derating

According to the curves in Fig. 4.8, if the loading or the temperature is reduced by 50%, the failure rate is reduced by one order of magnitude. This technique of operating an electronic device at less than its rated maximum capability is called *derating*. It is one of the key techniques for increasing reliability. The price of components rises, however, due to this type of overengineering.

As mentioned earlier, the exponential degradation caused by temperature has a significant impact on failure rates. If, for example, the operating temperature of integrated circuits is increased by 10 K, the average time to failure is approximately halved. Thus, the reliability of electronic systems is closely tied to their thermal management, especially cooling (Chap. 5).

4.4.5   Accuracy of Failure Rates

Failure rates for components are specified by manufacturers, users, and authorities. On the face of it, these specifications appear very convincing and are typically published with very accurate supporting data. However, in most cases, there is no

---

[3]Derating is the operation of an electronic device at less than its rated maximum capability in order to prolong its life.

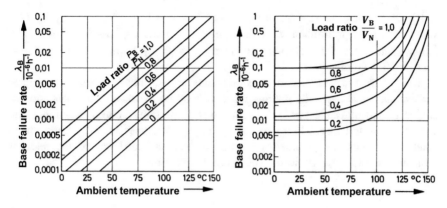

Fig. 4.8 Failure rate characteristics for carbon composition resistors (*left* [3]) and dry tantalum capacitors (*on the right* [4])

mention of functional and environmental stresses that components are subjected to during the observation period. For example, the simple statement "The failure rate of a film resistor is $33 \cdot 10^{-9}$ failures/hour" is unsatisfactory. The following data at a minimum should be provided as well:

- operational and test conditions: these include functional stresses, such as electrical and magnetic stress; and environmental stresses, such as ambient temperature, humidity, mechanical loading (such as vibrations), and the effects of radiation,
- limit values for failure criteria, including drift failure criteria,
- observation period and stress cycles, if necessary,
- confidence range for failure rate (i.e., the range that contains the true value with a given probability), calculated from the number of failures.

As these test conditions and requirements are typically not published, the literature often contains failure rates that differ by many orders of magnitude. However, comparative reliability and weak-point assessments can still be carried out with specifications from a *single* source. The same applies to concept tests.

Durability data for switches and relays cannot be specified directly, as they depend on the type of contact loading as well as environmental conditions (temperature, humidity, air impurities, shock, and vibration loading).

The typical service life spans of switches and relays in Table 4.3 are essentially "approved mechanical cycles." If, for example, the typical time to failure for a relay

Table 4.3 Typical life spans of switches and relays	Component	Number of cycles $\cdot 10^6$		
		Min.	Typical	Max.
	Pushbutton switch	0.01	0.1	10
	Relay, heavy-duty	1	2	10
	Relay, light-duty	1	10	100
	Reed relay, DIL enclosure	5	10	1000
	Reed contact, dry	0.5	50	5000

is $10 \cdot 10^6$ cycles, the equivalent failure rate is $\lambda = 0.1 \cdot 10^{-6}$ failures/cycle. This value should be multiplied with the cycles/time unit, e.g., 1000 cycles/h. The failure rate for the selected numbers $\lambda = 10^3$ cycles/h $\cdot\, 0.1 \cdot 10^6$ failures/ cycle $= 10^{-4}$ failures/h.

The failure rates of electronic components are essentially based on comprehensive statistical data, and nothing more. It is by no means a recommendation to engineers and other electronic designers to use this failure rate data without due consideration and care. More importantly, the set of curves for the component at hand, along with the parameters for temperature and loading and the relevant influence factors (π), will need to be carefully examined.

4.5   Failure of Electronic Systems

4.5.1   Calculation Principles

Before investigating the reliability of electronic systems, we first illustrate in this section the relationship between the reliability data for the components forming the system and the system itself. To do this, we generalize the analysis of the system and view its components and modules as system elements. The relationships introduced below will then be applied to actual systems in Sect. 4.6.

Information on the failure modes for system elements is required to predict system failure modes. The failure mode for system elements is typically calculated from the failure modes quoted in data sheets under reference conditions *and* considering the actual stress and environmental conditions (Sect. 4.4.2). The following condition must be met to establish the system failure mode from the element failure modes: *The functioning or failure of an element is independent of other elements, i.e., there is no mutual interaction of the reliability values.*

This criterion is met in many electronic systems. Hence, reliability calculations for electronic systems are much more useful than for mechanical systems, where failure rates for individual mechanical components cannot be accurately provided due to mutual, failure-related interactions between components.

Assuming the above condition holds, system reliability (e.g., of a printed circuit board) can be calculated from the failure modes of system elements (e.g., electronic components). The converse is also true, where failure modes for elements forming the system can be inferred from the system reliability. Both these approaches are based on the *serial* or *parallel structure of the reliability network* for the system introduced below in Sect. 4.5.2. It is very important that the relationship between the system and its reliability network model be understood before introducing the analytical techniques that are used to calculate the reliability of these systems in Sects. 4.6.3 and 4.6.4.

4.5.2 Network Modeling—Serial and Parallel Systems

A *structure* is said to be *serial* in the context of reliability if the entire system fails as a consequence of a failure of at least one of its constituent elements. For example, the electronic circuit on a printed circuit board typically fails when a component fails. The system failure rate is always greater than an individual element failure rate, and the probability of system survival is always less than that of an individual element. This illustrates the disadvantages of serial systems, in which high individual element reliabilities are required to achieve high system reliability.

A *parallel structure*, on the other hand, comprises a basic element and at least one standby element. Only one of the components in a parallel system needs to be working for the system to be operational. All components, that is, the basic element and all standby elements, must fail for the system to fail. Multi-engine aircraft, for example, can land safely with only one engine running. *Redundancy*[4] is defined as the availability of more technical functional elements than are required for the intended function. Standby elements constitute structural redundancy. The redundancy level for r elements is $r' = r - 1$. Therefore, a double-engine aircraft has a redundancy level of 1.

When all elements of a parallel configuration are active (switched on) and always connected to a working system this is called *hot-spare* or *hot-standby redundancy*. In the case of *cold redundancy,* the secondary or back-up element is activated (switched on) only when the primary element fails.

Redundancy greatly increases system reliability, as the probability of system survival in parallel configurations is higher than the survival rate of its individual components. However, this applies only up to the mean time to failure (*MTTF*) of individual redundant elements; the entire system can only be configured as reliable by means of redundancy up to the *MTTF* of its constituent elements (Sect. 4.6.4; Fig. 4.11).

Please note that the reliability network as a serial or parallel structure should not be mistaken for the real circuit layout. If, for instance, one of two "series" connected blocking capacitors fails due to a short circuit, the system can still work properly. This is technically a serial system but acts reliability wise as a redundant parallel system.

Generally speaking, the reliability network model depends not only on the functional configuration, but also on the type of failure. If the capacitor in the example above did not fail by short circuit, but by an open-circuit, such as a broken wire, the system would fail since the other blocking capacitor would become disconnected; this would then be a serial system in terms of reliability.

---

[4]The word redundancy comes from the Latin word *redundare*, which means *available in abundance*.

4.6 Reliability Analysis of Electronic Systems

4.6.1 Preliminaries

The systems treated below are assumed to be free of design, assembly, and man-ufacturing failures and do not show any effects of wear and tear. This applies to both hardware and software. Failures occurring after the system has been put in operation are exclusively random, and, hence, the probability theories based on the exponential distribution covered in Sect. 4.3.2 can be applied.

We should point out, however, that the reliability of a system should not be used for warranty purposes. This is due to inaccurate failure rates rather than reliability theory. Depending on the source of the failure rates, "favorable" or "unfavorable" numerical values can be chosen, resulting in widely differing reliability parameters. It is not uncommon to see failure rates that differ by orders of magnitude between various sources.

Reliability analyses are still useful, as they provide the development engineer with useful information for selecting and designing components, and for identifying weak points. Different system designs can be compared with respect to reliability. When doing so, the failure rates need to be taken from the same source and uniformly applied. Failure rates from different sources, or failure rates that are applied without specifying the functional and environmental stresses they were subjected to when determined, are useless.

Sales people should refer to these issues when, for example, a customer points to a better *MTBF* value from a rival bidder. Due to the widespread misunderstanding of *MTBF* as "average" or "useful life," engineers and sales should not use those values when talking to customers, nor should they promise failure-free life based on those values. A customer should stipulate the failure rates to be used when com-paring different quotes.

The types of reliabilities found in industrial practice are nominal, test, design, and operational reliabilities. As reliability should be ideally "designed in," the following discussion is based on *design for reliability*. This term is typically (and here as well) defined as "a reliability value that is calculated mathematically and is based on the design and the parts used, excluding manufacturing failures."

Design reliability should be calculated early in the design process, as it can provide valuable indicators of weak points and thus impact design decisions, such as the choice of components. These reliability calculations should be done con-currently with the development work, so that the development engineer has up-to-date information on the expected reliability of the system.

There is no real "lifetime" metric for electronic systems as every failure can technically be cleared. A system can be in service indefinitely depending on the strategy adopted by the organization. In practice, the end of the useful life of a system is reached when troubleshooting or operating the system becomes too expensive.

The term *minimum lifetime* should not be used for electronic systems. A minimum lifetime (sometimes referred to as *minimum service life* or *incubation period*) is the period in which a product never fails. This postulates that the probability of a failure is exactly (and not only approximately) zero. Such a period exists for certain items, like solder joints. Here, a series of time-consuming processes must take place before solder joints fail. However, it is impossible in practice to quote a minimum lifetime for an entire system as all individual components including their interactions would be required to possess this property.

The term *ROCOF* (rate of occurrence of failures) is often used for repairable systems. Adding the term *availability* to the parameters yields a useful practical reliability metric, as we show next.

4.6.2 Availability of Repairable Systems

We have already mentioned that the lifetime of systems can be extended infinitely by repair. The relationship between service period and out-of-service period (downtime) is of interest here. If the system is not working, it may have failed or be shutdown. Typically, the out-of-service period is set to the mean time between failure and repair, whereby the mean value for many repairs should be used. The probability of survival (reliability function) considers no repairs. It is worth examining how available systems really are, especially systems that are in the continuous operation.

The *availability A* is a performance criterion for repairable systems that accounts for both the reliability and maintainability properties. It describes the probability that a component or system will operate satisfactorily at a given point in time when used under stated conditions. The exact definition of availability is based on the types of downtimes one chooses to consider in the analysis. *Inherent availability* considers only repair downtime, expressed as mean time to repair (*MTTR*); it can be calculated with:

$$A = \frac{MTBF}{MTBF + MTTR} = \frac{1}{1 + \frac{MTTR}{MTBF}}. \tag{4.15}$$

If preventive maintenance downtimes are also included (in addition to corrective repair downtimes), the term *achieved availability* is used. Finally, the *operational availability* is a measure of the average availability that includes all experienced sources of downtime, i.e., administrative or waiting downtime, and logistic downtime, in addition to both preventive maintenance and corrective repair downtimes.

If we run the numbers with an example, we can get better insight into these metrics. A system has a mean time between failures *MTBF* = 1000 h and a mean time to repair *MTTR* = 10 h. The probability of survival $R(t)$ for the period of 1000 h is as follows:

$$R(1000 \text{ h}) = e^{-\lambda t} = e^{-t/MTBF} = e^{-1000 \text{ h}/1000 \text{ h}} \approx 0.37 \approx 37\%.$$

If the component is repaired, the inherent availability A is as follows:

$$A = \frac{1}{1 + \frac{10\,\text{h}}{1000\,\text{h}}} = 0.99 = 99\%.$$

Equation (4.15) shows that the same inherent availability is obtained for different mean failure intervals, when *MTTR/MTBF* is constant. Even unreliable systems can have a high permanent availability, if the out-of-service period or repair duration for failed components or modules is minimized. This illustrates the importance of a carefully considered and planned modular system configuration, good test facilities, and an efficient and capable repair service.

4.6.3 Electronic Systems Without Redundancy—Serial Systems

System reliability, as explained in Sect. 4.5, depends on the reliability of individual components and their interactions. Electronic systems without redundancy are systems in which all components must work to ensure system success, i.e., serial systems from a reliability point of view. In other words, if a single component fails, then the entire system will fail. We assume constant failure rates, i.e., no functional degradation, to simplify the analysis. We also assume that the reliabilities of individual components are known under their actual operating conditions, i.e., their failure rates λ_B were calculated under their actual operating conditions based on their FIT values (λ_{ref}) under reference conditions and adjusted using influence factors (π) (Sect. 4.4.2).

If the survival probabilities (reliability parameters) of single elements 1 to n in a system are $R_1(t)$ to $R_n(t)$, the system reliability $R_S(t)$ of the whole unit can be expressed as follows:

$$R_S(t) = R_1(t) \cdot R_2(t) \cdot R_3(t) \cdot \ldots \cdot R_n(t). \tag{4.16a}$$

We can see from this *product rule of reliability* that the probability of survival of a serial system is always less than the smallest individual probability of survival. The more individual elements in an electronic system, the more likely it is to fail. If, for example, the probability of survival of an individual component is 0.99 (99%) and if the system comprises 10 such components, the system has a probability of survival of $0.99^{10} = 0.904$, which is still >90%. If, however, the number of such components is 1000, the probability goes down to $0.99^{1000} = 0.00004$, i.e., 0.004%. So the fact that modern microprocessors with 5 million times more components, that is, 5 billion transistors, work at all is a testament to the high reliability of today's microelectronic components.

The product rule of reliability can be used not only to determine the system reliability, but also to determine the minimum required reliability of each component in order to satisfy an expected system reliability. For example, if the system reliability must not be less than 0.99, what is the minimum reliability of its 100 identical components? Using Eq. (4.16a), we get $0.99 = R_n(t)^{100}$, i.e., $R_n(t) = 0.99^{1/100} = 0.9999$.

Substituting the reliability function for the exponential distribution as per Eq. (4.8) in the product rule in Eq. (4.16a) yields:

$$R_S(t) = e^{-\lambda_1 t} \cdot e^{-\lambda_2 t} \cdot e^{-\lambda_3 t} \cdot \ldots \cdot e^{-\lambda_n t}, \qquad (4.16b)$$

$$R_S(t) = e^{-(\lambda_1 + \lambda_2 + \lambda_3 + \cdots + \lambda_n)t}. \qquad (4.16c)$$

The total failure rate of the serial system λ_S is thus the summation of the failure rates of all components:

$$\lambda_S = \lambda_1 + \lambda_2 + \lambda_3 + \cdots + \lambda_n. \qquad (4.17)$$

Assuming an exponential distribution, the mean time between failures (*MTBF*) or the mean time to failure (*MTTF*) may be calculated from the reciprocal of the failure rate (Eq. (4.12)). The *MTBF* for a serial system can then be expressed as follows:

$$MTBF_S = m_S = \frac{1}{\lambda_S}. \qquad (4.18)$$

If the system is subjected to environmental and operational stress during rest periods, then the failure rate is also applicable here and, hence, these periods should be included in the calculations. This scenario can occur in systems that are also partially operational during rest periods. Modules that have been switched off may be thermally impacted by heat, for example. The following formula can be used in such cases:

$$R(t) = e^{-(\lambda_{on} \cdot t_{on} + \lambda_{off} \cdot t_{off})}, \qquad (4.19)$$

where λ_{on} is the failure rate for the "on" state, λ_{off} is the failure rate for the "off" state, t_{on} the service time or "on" period, and t_{off} is the idle time or "off" period.

The procedure described above delivers only a useful approximate value when the system being monitored continuously executes all states, i.e., all failure options are effective and all components are active. The result of such a calculation can be too pessimistic with standard modules for logic circuits, where often only a subset of the available functions are used and failures in inactive elements have no effect.

Redundancies should be included in the calculations (see Sect. 4.6.4) if the system can only be partially viewed as a serial system in terms of its reliability; this can improve upon the "worst-case" approach taken here, whose results are therefore too pessimistic.

4.6.4 Electronic Systems With Redundancy—Parallel Systems

Redundancy techniques must be used if the desired design reliability cannot be achieved despite careful circuit design and the use of particularly reliable components. *Redundancy*, as explained in Sect. 4.5.2, is the deployment of additional components or modules that are not required for the main function (Fig. 4.9). The purpose of these redundant components is to increase the reliability, as they can perform the function of a failed element. Redundant components employed for the purpose of reliability are characteristic of parallel systems that comprise a basic unit and at least one standby unit.

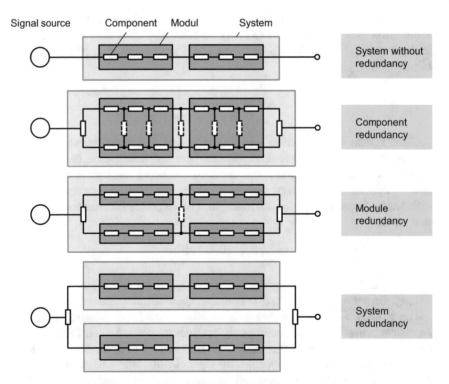

Fig. 4.9 Communication networks with surplus, and thus redundant, components. Redundancy can be built in at different levels, i.e., at the component, module, or system level

A number of parallel paths should be made available to the signal, so that it can follow an "alternative" route if an element fails. If, for example, there is a switch in a transmission line that does not reliably *close*, the signal can be shunted through other switches to mitigate the unreliability effects. The same applies to switches that do not reliably *open*; other switches can be arranged in series in this case. Unreliable opening and closing can be mitigated with a combination of four switches in a serial–parallel cluster. This "quad redundancy" principle is shown in Fig. 4.10 with four diodes.

Redundancy can be built in at the lowest level, that is, in the components (with a suitable parallel component in the simplest case, for example); at a higher level in the modules; or at the system level (see Fig. 4.9). Increasing the redundancy level at the component level for *all* components in a system leads to a greater increase in reliability than a comparable increase in redundancy at the system level. To put it more simply, duplicating all components in a system effects higher reliability for the user than duplicating the system.

Redundancy elements are called "active" or "hot" if they are always operational during the service period, as described in Sect. 4.5.2. The concepts of hot-spare or hot redundancy are relevant in this context. Redundancy elements are "passive" or "cold" if they come into operation only when they are actually needed. A system that detects a faulty component and actuates a control element is required for this type of cold standby or cold redundancy. The reliability of this failure monitor and the changeover switch must be considered as well and integrated in the required level of reliability for the parallel circuit. These complex cold standby redundancy issues are beyond the scope of this book; please refer to [7] for a detailed treatment of this topic.

The probability of survival (reliability function) of a redundant system with hot-spare redundancy is calculated from the failure probabilities of the same or similar redundant elements, as the system fails exactly when all elements fail in the given period. If $F_1(t)$ to $F_r(t)$ are the failure distribution functions of redundant elements 1 to r in a system with redundancy level $r - 1$, the system failure function $F_S(t)$ can be expressed as follows:

$$F_S(t) = F_1(t) \cdot F_2(t) \cdot F_3(t) \cdot \ldots \cdot F_r(t). \tag{4.20}$$

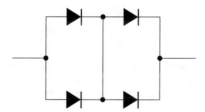

Fig. 4.10 Diode circuit with quad redundancy, in which the outputs of one stage are interconnected to the inputs of the succeeding stage by a connection pattern so that failures in the first stage are overridden in the second stage

This *product rule of unreliabilities* shows that the failure probability of a parallel system is always smaller than the smallest individual failure probability (failure distribution function) of its elements. Should the theoretical assumption hold that a single element i cannot fail ($F_i(t) = 0$), then the parallel system will not fail either ($F_S(t) = 0$ or $R_S(t) = 1$).

Obviously, the unreliability of the system decreases as the number of parallel components is increased; hence, the reliability increases with the number of redundant components. Figure 4.11 illustrates the possibilities and limitations of this increase in reliability with a parallel configuration of redundant elements. On the one hand, the *respective* increase in the probability of survival drops with increasing redundancy level: This means that single-digit redundancy levels should be used. On the other hand, a significant increase in reliability by means of redundancy is observed only in the period of *MTBF* (or *MTTF*) of its elements. The probability of survival for this period can be increased from 37% (no redundancy) to almost 100% with a redundant system, but then the probability of survival will inevitably drop. This is due to the fact that a system with redundancies cannot remain reliable for a longer period than its constituent elements.

Finally, it is important to be aware that simplifications during the investigation of the reliability of redundant systems can lead to false conclusions. Critical analysis is recommended, especially in connection with possible failure modes, as they can impact the changeover to the standby unit and call into question the correct assumption of a parallel structure in the context of reliability.

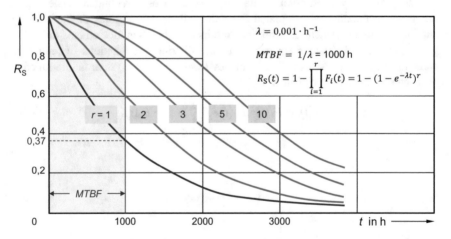

Fig. 4.11 Increase in the probability of survival of a system with hot-spare redundancy for redundancy levels $r - 1 = 0, 1, 2, 4, 9$. The graph shows the decreasing benefit of ever increasing redundancy levels as well as the reliability gain limited to the period of the mean time between failures (*MTBF*) of its elements

4.6.5   Service and Maintenance of Electronic Systems

A system is *serviced* to keep it in operation. Servicing can be divided into the three areas: inspection, maintenance, and repair. *Maintenance*, in contrast to repair, is a preventative measure to counteract wear and tear.

The random failures typically encountered in electronic systems cannot be prevented by maintenance. A sudden short circuit or an interruption at normal load is not due to wearout and are thus not predictable. *Drift fails*, i.e., a slow change in attributes, are easier to deal with. Drift, which typically cannot be detected in digital operating systems, causes output voltage shifts, such as change in gain amplification, in analog components. A total failure can be prevented here with regular inspections and maintenance, especially by monitoring, tuning, and replacing components when necessary. Drift-related operation and failure modes for non-redundant systems with and without maintenance are shown in Fig. 4.12.

The standby system takes over operation in the event of a failure for redundant systems. If the failure is found during equipment condition assessment or maintenance, a failure of the standby system has no impact.

Components should be replaced during planned maintenance if they exceed their expected (intrinsic-fails-only) lifetime and, hence, are in danger of becoming subjected to *wearout failures*. This applies especially to electromechanical components with high cyclical or continuous loading, for example, fans for reducing the

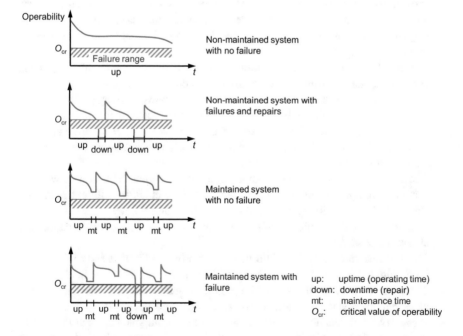

Fig. 4.12 Drift characteristics of non-redundant maintained and non-maintained systems

internal system temperatures. The wearout-failure distributions should be considered when designing systems with such components to enable preventive maintenance schedules that are time-coordinated for all components. Determining the times at which expendable modules are to be replaced should be a crucial part of early design decisions.

4.7 Recommendations for Improving Reliability of Electronic Systems

All employees involved in the design, development, and manufacture of reliable systems should be proactively engaged with reliability issues. A range of measures are required to assure the high reliability standards of components and systems are met. The key considerations are outlined below. All of these measures have associated material or personnel costs. A cost-benefit analysis is required to determine an appropriate balance:

- The lowest number of components possible should be used in a given solution. Please note that the failure rates for different components can vary greatly. The total failure rate, for example, of 100 carbon-film resistors is lower than that of one-film potentiometer. A potentiometer, therefore, should be replaced by a voltage divider made of two resistors.
- Proven, tried, standard components should be used. Caution is advised with new and unknown components. A balance needs to be struck between the benefits of improved functionality (e.g., using the latest VLSI circuits) and the risk of unknown reliability.
- Use reliable component types: failure rates can be greatly reduced by using ceramic capacitors instead of electrolytic capacitors, for example.
- Operate components below their nominal ratings (power, voltage, current, etc.). Operating components only at 50% power rating reduces the failure rate of many components to a fifth, and sometimes up to a tenth, of its original value. The cost of such overdesigned components increases in many cases approximately proportional to the rated load.
- The various components and modules of a system should be designed with the same reliability. Designing different elements in a chain to different reliability standards is, generally speaking, a waste of time and money.
- Components and modules should be operated at ambient temperatures that are as low as possible.
- Components and modules should be prematurely aged so that early failures can be eliminated before the components and modules leave the test field.
- Components and modules should be protected against climatic, mechanical, and electrical stresses during storage, transportation, and operation.
- Preventive maintenance should be carried out if components reach the end of their expected (intrinsic-fails-only) lifetimes. Wearout-failure distributions

should be considered in the design process to enable time-coordinated maintenance schedules of expendable components and modules.

- Elements that are prone to failure or are subject to wear should be easy to service and repair, and, hence, easy to access. These requirements should have a higher priority than the typical desire for high packaging densities.
- Redundancy techniques should be applied where there is insufficient reliability. Higher system reliability can be achieved by structural redundancy at the component level than with similar levels of redundancy in higher levels of the system. However, an increase in system reliability by means of redundancy is only achievable during the *MTBF* (or *MTTF*) period of its redundant elements; reliability inevitably drops after this period has elapsed. Hot-spare redundancy avoids failure monitoring that is prone to error, and no changeover mechanism is required.

References

1. IEC 61709:2011 *Electric components - Reliability - Reference conditions for failure rates and stress models for conversion*, publication date 2011-06-24
2. *Military Handbook Reliability Prediction of Electronic Equipment*, MIL-HDBK-217F, 2 December 1991
3. MIL-R-11/RC 07/12: *Electronic Components - Resistors Military Ordering Code*
4. MIL-C-26655: *Electronic Components - Capacitors Military Ordering Code*
5. SN 29500: *Failure Rates for Electronics and Electromechanical Components.* Siemens AG Munich, Corporate Technology, Dept. CT TIM IR SI
6. ISO 13849-1:2015, *Safety of machinery – Safety-related parts of control systems – Part 1: General principles for design*, December 2015
7. A. Birolini, *Reliability Engineering. Theory and Practice*, 7th edition, Springer, 2014

Chapter 5
Thermal Management and Cooling

A considerable amount of heat energy is typically generated in electronic systems due to power dissipation: When current flows through a resistor, heat is generated by the friction between electrons moving against one another. This leads to thermal stress with a negative impact on operation and reliability. Technological advances in electronic modules allow ever higher packing densities and switching frequencies, which give rise to a dramatic increase in the amount of heat generated per unit volume. Thermal constraints are becoming more critical as well in the design and development of electromechanical and mechanical assemblies, owing to the increasing trend to more compact designs and higher power conversion per unit volume. The problem is further exasperated by the increasing use of poor heat-conducting plastics.

We showed in Chap. 4 that temperature is one of the most important factors influencing component failure rates. It is known that cutting the operating temperature (measured in degrees centigrade) of a component by 50% can result in a five- to tenfold reduction in its failure rate. Also, the failure rate of semiconductor devices is roughly halved for each 10 degrees centigrade reduction in junction temperature.

For these reasons, thermal constraints must be considered carefully when designing electronic systems. The approved temperatures need to be observed throughout the entire product life cycle under the expected operating conditions. The thermal design must ensure that temperatures within the system always remain within the specified limits.

This chapter will prepare you for determining the heat energy produced by a system (Sect. 5.1) and for calculating the heat paths with thermal networks (Sect. 5.2). The heat transfer principles described in Sect. 5.3 will help you select and design suitable elements for heat dissipation (Sect. 5.4). They assure compliance with the thermal requirements for real-world electronic systems (Sect. 5.5). Finally, recommendations for thermally correct systems design will be given in Sect. 5.6.

© Springer International Publishing AG 2017
J. Lienig and H. Bruemmer, *Fundamentals of Electronic Systems Design*,
DOI 10.1007/978-3-319-55840-0_5

5.1 Introduction—Terminology, Temperatures, and Power Dissipation

5.1.1 Problem Definition

Energy losses in electronic systems, e.g., in resistors or semiconductors at the component level, generate heat energy. This heat is removed from the heat source at the heat transfer rate $\dot{Q}$, i.e., heat energy per time unit. However, the heat removal is often incomplete, which can cause a considerable temperature rise in the system.

In the case of most information processing electronic components and systems, only a very small portion of the supplied electrical power goes to information processing. Most of it is dissipated as heat energy. In fact, we can safely assume that almost 100% of the electrical energy supplied to an electronic component or system is converted to heat and thus wasted. This heat not only damages electronic components, but also increases the signal noise in semiconductor devices due to an increase in the movement of free electrons.

The more efficiently the excess heat is removed, the lower the temperature rise in the component and in the system. For example, it is crucial to hold the junction temperature of a semiconductor device below the maximum temperature specified by its manufacturer. Hence, the development engineer is tasked with selecting suitable heat removal mechanisms and implementing them with effective design measures. Heat can be removed by three transfer modes: conduction, convection, and radiation.

As an introduction to these issues, we will examine the removal of the heat caused by power dissipation P_D from a resistor R located in a sealed enclosure (Fig. 5.1). We assume that the heat transfer is neither promoted nor obstructed by the mounting method for the resistor.

Applying a current I through a resistor R causes a voltage drop V across the device. The power dissipation P_D of the resistor arises from the input electrical energy P and is given by:

$$P_D \approx \frac{V^2}{R} = I^2 \cdot R. \tag{5.1}$$

Power dissipation causes a temperature rise ΔT in the resistor with respect to the surrounding air. This temperature difference leads to a heat transfer rate $\dot{Q}$ between the resistor surface and the environment. The resistor temperature rises with respect to the temperature of its surroundings until the electrically generated heat loss P_D equals the released heat transfer rate $\dot{Q}$. When the power and thus the heat loss P_D are suspended, the resistor cools down and the temperature difference between the component and its surroundings disappears.

We assume a constant temperature, the *operating temperature*, in steady-state operating conditions where the heat generated equals the heat removed by the

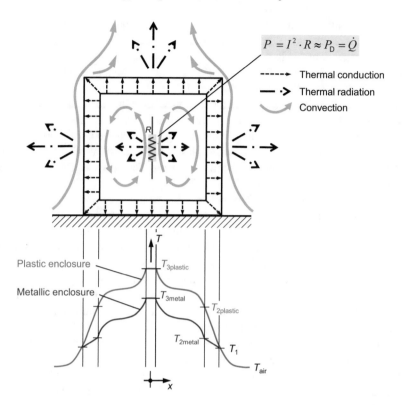

Fig. 5.1 Illustration of the power dissipation P_D from a resistor arising from the input electrical energy P in an enclosure (*top*). The associated temperature distribution is also shown (*bottom*), indicating that lower thermal resistances between the component and the outside world across, for example, a metallic enclosure reduce the resistor's temperature

cooling mechanism $\left(P_D = \dot{Q}\right)$. This situation is preferred for thermal calculations, as the mathematics for heating and cooling are challenging.

How is the heat dissipated from the surface of the resistor in Fig. 5.1? The heat contained in an object, a liquid, or a gas depends on the motion of its molecules. This motion increases as the heat increases. The *temperature* is the metric for the instantaneous mean value of the kinetic energy of the molecules. If you place a warm object in a medium that is at a lower temperature, such as air, the moving molecules in the hotter object transmit on average more kinetic energy by elastic collisions to the molecules in the colder medium than vice versa. This process continues until a new common mean temperature is reached, that is, until the *temperatures of both mediums have been equalized*. This heat transfer process is irreversible, as we know from the second law of thermodynamics.

The vibrating molecules on the surface of the resistor in Fig. 5.1 transfer their energy to the molecules in the surrounding air. The heated air in the vicinity of the resistor expands. Its density is therefore lower (it is "lighter") than air that is further away. The resulting uplift produces an airflow inside the enclosure. Heat is transferred to the enclosure when the airstream comes in contact with the colder enclosure panel. The air is then cooled, and it becomes "heavy" and "falls" to the bottom of the enclosure. This process is called *convection* (Sect. 5.3.3) and is illustrated by the circular arrows in Fig. 5.1.

In our example, the heat is next transported by vibrations from molecule to molecule and further into the enclosure panel. This effect called *thermal conduction* (Sect. 5.3.2) can occur in solids, liquids, and gases. The medium is key here, for example, the enclosure panel material. A plastic enclosure "resists" the heat transfer to a much greater extent than a metallic enclosure, for instance. Convection takes place again on the panel exterior.

There will be no passive airflow between the heat source and panel if the space between resistor and enclosure panel is very tight (some few millimeters). The heat transfer in this case is almost entirely by thermal conduction. The almost stationary air acts as a considerable barrier to the heat transfer. This phenomenon is used for good effect, for example, in insulated glass windows, where there is a narrow strip of air between two glass panes and reducing heat transfer is the goal.

Heat transfer through *radiation* also occurs in the example in Fig. 5.1 (top) along with physical convection and thermal conduction. Thermal radiation (Sect. 5.3.4) is an electromagnetic oscillation that is propagated according to the laws of optics. Thermal radiation is an exchange of radiation between the surfaces of objects. Air is not heated by radiative transfer as it is practically fully transmissive for thermal radiation. Thermal radiation propagates linearly and is absorbed, reflected by, or passes through matter. In the case of absorption, it is reconverted to heat (molecular excitation).

There is nearly always a combination of thermal conduction, convection, and radiation occurring in electronic systems. However, they are handled differently in the thermal calculations as they obey different physical laws.

Finally, we will describe the *temperature distribution* between the "heat source", resistor R, and the ambient air outside the enclosure (Fig. 5.1 bottom). To simplify matters, we assume uniform heat dissipation through the enclosure in these analyses. Working from the outside of the enclosure, in toward the resistor, we can see the following:

- The air temperature outside the enclosure (T_{air}) is the lowest temperature. The temperature rises due to convection in a nonlinear fashion to the external surface temperature T_1 of the enclosure. As indicated in Fig. 5.1 (and explained in Sects. 5.5.2 and 5.5.5), *this external enclosure temperature is independent of the inside temperature of the enclosure*. It depends only on the heat to be transferred and the thermal resistance between enclosure and ambient.
- The temperature differential in the enclosure panel is a function of the thermal conductivity of the panel material. A poor heat conductor requires a higher temperature difference ($T_{2plastic} - T_1$) for a given heat dissipation P_D to pass through than a good conductive material ($T_{2metal} - T_1$).

- The air temperature inside the enclosure depends on the heat transfer of the heated air to the inside of the enclosure panel. This relationship is also nonlinear due to convection.
- The surface temperatures of the resistor ($T_{3\text{metal}}$ and $T_{3\text{plastic}}$) differ significantly in metallic and plastic enclosures. Note that the "shape" of the curve between the resistor and the enclosure is very similar for both the plastic and metal enclosures, and that the increased temperature for the plastic enclosure curve results from the increase in temperature produced by thermal conduction across the enclosure.

Overall, the reduction in thermal resistances between the component and the outside air temperature is the key to keeping the component's temperature below its maximum value.

The processes covered only qualitatively here by way of introduction will be more closely examined below, including definitions of the most important basic concepts. Section 5.2 deals with the engineering approach to thermal management. An analogous model, which uses a thermal network equivalent to an electrical circuit, is used to calculate the heat transfer. Section 5.3 covers the physical fundamentals of heat transfer. This knowledge is necessary for understanding the underlying relationships and for correctly selecting and putting in place effective heat dissipation strategies. Section 5.4 describes critical heat dissipation elements used to solve thermal problems because of their beneficial heat transfer properties. Their application in the field of electronic systems design is dealt with in Sect. 5.5.

5.1.2 Important Parameters in Thermal Management

If an electric potential difference V is applied to a resistor R for time t or if a current I flows in the resistor, the electrical work

$$W = V \cdot I \cdot t = \frac{V^2}{R} \cdot t = I^2 \cdot R \cdot t = P \cdot t \tag{5.2}$$

is performed; the *electrical energy* W is converted to the same amount of *thermal energy*. The thermal energy of an object is manifested in its temperature and equals the mean kinetic energy of its molecules.

Energy is the ability of a physical system to do work, for example, the ability to cause an electrical current to flow through a resistor. There are many forms of energy: mechanical, electrical, magnetic, chemical, thermal, etc. According to the law of conservation of energy, energy cannot be created or destroyed, it can only be converted.

A temperature difference between two adjacent systems or objects produces an energy transfer; energy is transferred from the high-temperature system to the low temperature one. Thermal energy that is transported across system boundaries is called *heat energy Q* (also: *quantity of heat* or simply *heat*). It defines how much thermal energy one object transfers to another object. As an object dissipates heat energy, its thermal energy is reduced and vice versa. Heat energy, in accordance with the second law of thermodynamics, is always transmitted from a system at a higher temperature to a system at a lower temperature. Heat energy characterizes the process of transmitting thermal energy from one object to another or also from an object to its environment. It is therefore a process variable. The unit of heat energy is joule (J).

If we are interested in learning how quickly heat is transferred between two objects, we need to investigate the heat transfer per unit time. This is called the *heat transfer rate*, also known as *heat transfer* or *heat flow*, denoted by $\dot{Q}$ (or q). Hence, the heat transfer $\dot{Q}$ is the heat energy Q transported per unit time. It is thus an output (*heat output*) and has the unit watt (W) or joule per second (J/s). Note that heat transfer rate and power are the same physical quantities.

In many practical applications, it is important to know the density of this heat transfer rate, i.e., the heat transfer per unit area. The *heat flux density* or *heat flux* $\dot{q}$ (or $\overrightarrow{\Phi}_q$) describes the heat transfer $\dot{Q}$ through the surface area A. The unit for heat flux density is W/m^2 or $J/(m^2s)$.

Let us now apply these physical quantities to electronics. Heat flow in electronic components or systems is based on their respective *power dissipation* or *heat loss* P_D. This is the difference between the energy supplied to an electrical component or system and its desired released energy. It is therefore an undesirable portion of the energy converted to a heat loss in a component or system. The unit of power dissipation is therefore the same as that of heat transfer, namely watt (W). For an electrical system with no "effective output" to ambient, such as analog measuring instruments, the input electrical energy is fully converted to heat Q and completely dissipated as a heat transfer $\dot{Q}$ in the steady-state scenario (Fig. 5.2).

The *density of heat to be dissipated*, which corresponds to the heat flux density $\dot{q}$ introduced earlier, is a key parameter for electronic components and systems. Here, the power dissipation is related to the respective surface area of a chip or wiring substrate, and the unit used is W/m^2. The power to be dissipated is sometimes cited with respect to the volume of components or systems; typical units are here W/m^3 or W/dm^3.

The *temperature T* describes the thermal state of an object. It is a measure for the mean kinetic energy of the particles in a material. The temperature is defined in kelvin (K) based on absolute zero or, in degrees centigrade (°C), linked with the melting point of ice; temperature differentials are always quoted in kelvin.

There is no molecular motion at -273.15 °C. This value represents *absolute zero* and is the origin of the Kelvin temperature scale. The temperature interval "one degree" is the same on the Celsius and Kelvin scales. The latter is defined in the International System of Units (SI system) as a thermodynamic temperature scale with the unit kelvin as it simplifies the formulation of thermodynamic laws; it is

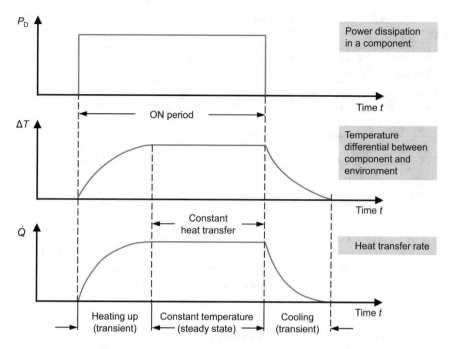

Fig. 5.2 Power dissipation, temperature, and heat transfer in an electronic component

represented by the symbol T. The symbol ϑ is sometimes used for the centigrade temperature scale. The following formula converts between the Kelvin and centigrade scales:

$$\frac{T_K}{K} = 273.15 + \frac{T_C}{°C}. \tag{5.3}$$

Equation (5.3) is illustrated with the following example:

$$T_C = 40\,°C \; (= 104\,F), \quad \frac{T_K}{K} = 273.15 + \frac{40\,°C}{°C}, \quad T_K = 313.15\,K.$$

When heat passes through an object, a temperature differential is formed along the heat flow path. The relation between this temperature differential (temperature drop) and the heat transfer is the *thermal resistance* R_{th} of the path. This is analogous to the electrical resistance R, which is calculated as the ratio of the electric potential difference across a component and the current flowing through it. Similarly, the thermal resistances of a multi-layer panel are added together like electrical resistors connected in series. The thermal resistance is used to characterize heat sinks. Its unit is K/W.

If heat flows into an object, its temperature increases. The relationship between input heat and temperature rise is called the *thermal capacity* C_{th}; it is measured in

Table 5.1 Main physical quantities in thermal management

Physical quantity	Symbol	Unit
Heat energy, heat, quantity of heat	Q	J
Heat transfer rate, heat flow	$\dot{Q}, q$	W
Power dissipation, heat dissipation, heat loss	P_D	W
Heat flux density, heat flux	$\dot{q}, \overrightarrow{\Phi}_q$	W/m^2 (W/m^3)
Temperature	T	K, °C
Thermal resistance	R_{th}	K/W
Thermal capacity, heat capacity	C_{th}	J/K or W·s/K
Thermal conductivity	k	W/mK
Heat transfer coefficient (convection, radiation)	h	W/m^{2}K

J/K or (W·s)/K. The thermal capacity (or heat capacity) is thus the ability of an object to store thermal energy. There is an analogy between thermal capacity and electrical capacitance C, which defines the storage capacity for electrical energy.

Table 5.1 summarizes the main parameters in thermal management.

5.1.3 Temperatures of Components and Systems

Increasing the temperature of an electronic component or system degrades its reliability. Approved ambient temperatures must be defined and designed to, in order to assure component and system reliability.

The *operating ambient temperature range* applies during operation, and the *transit ambient temperature range* applies in the "off" state, e.g., during assembly, storage, or transportation. Please note that the transit temperature range is typically greater than the operating temperature range. For example, the transit ambient temperature range for many laboratory instruments is −40 to 60 °C (−40 to 160 °F), while the operating ambient temperature range lies between 10 and 40 °C (50 to 104 °F).

Electronic components act as internal heat sources due to power dissipation; hence, the internal system temperature is higher than the ambient temperature. The surface temperature of components should always be used to determine the operating ambient temperature range for such systems, even if the components themselves do not produce any heat.

The temperature differential between a component or system that heats up internally and a defined ambient temperature is called *overtemperature*. A *temperature limit*, on the other hand, is defined as the maximum approved absolute temperature. In silicon-based semiconductor devices, for example, the junction temperature is limited to 125 °C (257 °F) for safe operation.

Approved overtemperatures for different applications of electronic systems are defined in standards and guidelines. Other temperature limits apply for system peripherals, e.g., controls, under health and safety regulations, e.g., risk of burns.

5.1.4 Power Dissipation in Electronic Components

All electronic components and wire interconnects dissipate heat in the form of heat flows $\dot{Q}$. The quantity of heat dissipated is expressed as the power dissipation P_D. Both line losses and frequency-dependent switching losses occur in electronic systems. For example, the power dissipated by a semiconductor or passive device equals its voltage drop multiplied by the applied current. The heat dissipation of a CMOS device depends on its frequency and geometry; here, switching power constitutes about 70–90% of the power dissipated.

By applying the theoretical assumption that almost all input electrical energy is converted to heat loss and thus to heat flow, thermal management always includes a high safety factor.

Table 5.2 gives an overview of the power dissipation calculations for important components and their determining parameters.

Table 5.2 Theoretical power dissipation P_D for selected electronic components [1]

Component	Power dissipation	Determining parameters
Resistor, conductor (interconnect)	Ohmic losses: $P_D = I^2 \cdot R = I^2 \cdot \rho \cdot \frac{L}{A} = I^2 \cdot \frac{1}{\sigma} \cdot \frac{L}{A}$	I line current R conductor resistance ρ specific electrical resistance σ specific electrical conductance L conductor length A conductor cross section
Capacitor	Dielectric losses with harmonic AC voltage: $P_D = V^2 \cdot \omega \cdot C \cdot \tan \delta$	V r.m.s. value of capacitor voltage ω angular frequency C capacitance $\tan \delta$ dielectric loss angle
Diode	$P_D = V_d \cdot I_d$	V_d diode voltage I_d diode current
CMOS devices	Switching losses (70–90% of losses): $P_D = C \cdot V_{dd} \cdot f$	C load capacitance V_{dd} supply voltage f switching frequency
Bipolar junction transistor	$P_D = V_{CE} \cdot I_C + V_{BE} \cdot I_B \approx V_{CE} \cdot I_C$	V_{CE} collector–emitter voltage I_C collector current V_{BE} base–emitter voltage I_B Base current
Junction field effect transistor, JFET	$P_D = I_D^2 \cdot R_{DS\,(on)}$	I_D drain current $R_{DS(on)}$ drain–source resistance

5.2 Calculation Principles

5.2.1 Electrical and Thermal Networks

Thermal management for electronic systems aims to maintain the temperature of given component parts, such as the surface of a resistor or the barrier junction of a semiconductor, as low as possible. The surface temperature of the resistor R in Fig. 5.1 is said to be low when the power dissipation (the heat flow) to ambient meets as low a "resistance" as possible. Therefore, the *thermal resistance* between the surface of the resistor R and the external air at temperature T_{air} must be as low as possible.

The electrical engineer can use analogies in electrical circuits to calculate the heat flows for power dissipation and the resulting component temperatures. Temperatures can thus be calculated via thermal resistances in a "thermal circuit" in the same way as voltages are calculated via electrical resistances. The illustrative electrical circuit in Fig. 5.3 comprises two voltage sources V_1 and V_2 and a resistance R. According to Kirchhoff's second law, the sum of all voltages in a circuit is zero:

$$V_1 - V_2 - I \cdot R = 0, \quad V_1 - V_2 = I \cdot R. \tag{5.4}$$

Thus, the voltage difference can be calculated by multiplying the current by the resistance value.

An electric potential difference applied across a given resistor produces a specified current flow. If there are a number of voltage sources in a circuit, their sum (or difference) yields the overall potential difference.

We can model a thermal problem with an equivalent electrical circuit in the form of a *thermal network* with the following analogies:

- electric voltage $V \equiv$ temperature T,
- electrical resistance $R \equiv$ thermal resistance R_{th}, and
- electrical current $I \equiv$ heat flow $\dot{Q}$ (power dissipation P_D).

The temperature T_3 in the example in Fig. 5.3 is the surface temperature of resistor R in Fig. 5.1, and temperature T_{air} is the ambient air temperature outside

Fig. 5.3 Electrical circuit (*left*) and analog thermal network (*right*)

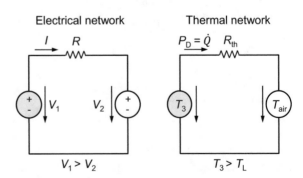

the system's enclosure. The thermal resistance R_{th} comprises the heat transfer "obstacles" described in previous sections. We can thus formulate Eq. (5.5a), similar to Eq. (5.4), as follows:

$$T_3 - T_{air} = P_D \cdot R_{th}. \tag{5.5a}$$

The temperature differential can thus be calculated by multiplying the power dissipation by the thermal resistance value. Hence, the surface temperature of resistor T_3 is the product of the power dissipation P_D and the thermal resistance R_{th} plus the ambient air temperature T_{air}:

$$T_3 = P_D \cdot R_{th} + T_{air}. \tag{5.5b}$$

The product "heat loss times thermal resistance" must be minimized to maintain a low resistor surface temperature T_3. We normally have little influence on the temperature T_{air} of the system's ambient air.

By rearranging Eq. (5.5b), we can calculate the permissible dissipated heat P_D for a known thermal resistance R_{th} and approved temperatures T_3 and T_{air}:

$$P_D = \frac{1}{R_{th}} \cdot (T_3 - T_{air}). \tag{5.5c}$$

Clearly, the transmitted power dissipation is proportional to the temperature differential across R_{th}.

We need to break down the thermal resistance R_{th} into its thermal conduction, convection, and radiation components to determine its size. Figure 5.4 shows these components for the example in Fig. 5.1. The individual temperatures that allow the respective thermal resistances to be determined are located at the nodes in the thermal network, similar to the voltages in an electrical network.

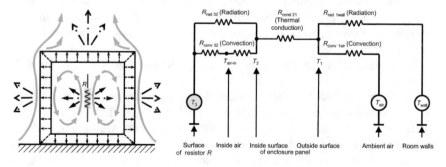

Fig. 5.4 Heat dissipation from a resistor in an enclosure (see Fig. 5.1) with corresponding thermal network comprising the three heat transfer mechanisms conduction, convection, and radiation

We can thus determine critical thermal parameters, such as surface temperatures and permissible power dissipation, by converting a thermal problem to a thermal network that is similar to an electrical circuit. This well-known design method, the *thermal network method*, is dealt with in the next Sect. 5.2.2.

5.2.2   Thermal Network Method

It is generally extremely difficult to mathematically model the complex thermal processes in an electronic system, due to the distributed nature of heat sources. A variety of techniques are available for thermal management, for example, calculating temperature fields in a system. We introduce an approach that is particularly suitable for electrical engineers, the thermal network method.

As discussed in Sect. 5.2.1, thermal processes can be modeled by an electrical circuit in the thermal network method, i.e., a network comprising discrete elements (supplies, resistors, capacitors). This is achieved by using analogy constants to model the similarities between a thermal and an electrical flow field. This analogy is very helpful as thermal and electrical quantities are practically identical. We can, for example, establish a correspondence between the flow quantities electrical current (i) and heat flow ($\dot{Q}$), as well as the two-point quantities electric potential difference (V) and temperature differential (ΔT).

We thus construct a thermal equivalent circuit that looks like an electrical network from the outside, but which models the thermal flow field in the system under examination. The rate of heat flow $\dot{Q}$, as already mentioned, is thus modeled by the electrical current i. The current i is generally the predefined quantity, as a given heat flow rate $\dot{Q}$ typically needs to be dissipated by the system. The individual temperature differentials ΔT correspond to potential differences V, which, as in electrical circuits, are determined by the size of the resistances causing them. We are dealing with thermal resistances R_{th} here, e.g., from the system interior to ambient through enclosure panels. These thermal resistances must be kept as small as possible to provide a small temperature differential and thus good heat transfer to ambient.

The corresponding quantities for electrical–thermal analogies are listed in Table 5.3.

The parameters listed in Table 5.4 and the elements depicted in Table 5.5 are used in thermal networks based on the analog relations from Table 5.3.

The network is constructed by defining the components, whose temperatures, based on a given ambient temperature, are to be calculated. These elements are depicted as *network nodes*. The individual component temperatures are calculated at nodal points.

The heat generated by the elements are represented as *heat sources*. *Thermal resistances* model the heat transfer processes in the system and are situated between

Table 5.3 Analogy between electrical and thermal quantities

Electrical	Thermal
$i(t) = C \cdot \dfrac{\mathrm{d}V}{\mathrm{d}t} + \dfrac{V}{R_{\mathrm{in}} + R_{\mathrm{out}}}$	$\dot{Q}(t) = C_{\mathrm{th}} \dfrac{\mathrm{d}(\Delta T)}{\mathrm{d}t} + \dfrac{\Delta T}{R_{\mathrm{th\_in}} + R_{\mathrm{th\_out}}}$
$i = \dfrac{V}{R_{\mathrm{in}} + R_{\mathrm{out}}}$	Steady state: Heat transfer $\dot{Q} = \dfrac{\Delta T}{R_{\mathrm{th\_in}} + R_{\mathrm{th\_out}}}$
Current i	in A	Heat transfer $\dot{Q}$	in W
Current density J	in A/m^2	Heat flux density $\dot{q}$	in W/m^2
Electric potential difference V	in V	Temperature differential ΔT	in K
Resistance $R = \frac{V}{i} = \frac{1}{\beta \cdot A}$ mit $\beta = \frac{\sigma}{L}$	in Ω, V/A	Thermal resistance $R_{\mathrm{th}} = \frac{\Delta T}{\dot{Q}} = \frac{1}{h \cdot A}$	in K/W
Capacitance $C = \frac{Q}{V}$	in A·s/V	Thermal capacity $C_{\mathrm{th}} = \frac{Q}{\Delta T} = c \cdot m$	in W·s/K
Spec. el. conductance σ	in A/(V m)	Thermal conductivity k	in W/(K m)

h Heat transfer coefficient in W/(K m^2)
c Specific heat capacity in J/(K kg) or W s/(K kg)
Q Heat energy, quantity of heat in J bzw. W s, electrical charge in A s
m Mass in kg

Table 5.4 Variables and parameters in electrical and thermal networks

Electrical network	Thermal network	Symbol
Current in A	Heat transfer in W	$\dot{Q}$
Voltage in V	Temperature in K	T
Charge in A·s	Heat energy in W·s	Q

the network nodes. The ambient temperature or any other fixed temperature is considered in the thermal network as a *temperature source*.[1]

---

[1]The temperature source in a thermal network is less a "source" and instead is more an element with a predefined temperature that cannot be impacted by the network. Temperature sources are integrated in the thermal network where constant or predefined temperatures are specified, for example, constant external (room) temperatures.

Table 5.5 Elements in thermal networks

Electrical network	Thermal network	Symbol
Ohmic resistance in V/A bzw. Ω	Thermal resistance in K/W	ΔT $\rightarrow \dot{Q}$ R_{th}
Current source in A	Heat source in W	$\rightarrow \dot{Q}$
Voltage source in V	Temperature source in K	ΔT + -
Capacitance in A s/V	Thermal capacity in J/K or W·s/K	ΔT $\rightarrow \dot{Q}$ C_{th}

There are two different types of network elements:

– If the steady-state temperatures or the temperature distribution of individual elements are to be calculated dependent on the input current to the system at constant ambient temperature, a network, made up of resistors, heat source(s), and temperature source(s), is sufficient for this *steady-state scenario*.
– If the *temperature distributions* at the individual nodes are to be calculated based on the input current and/or a changing ambient temperature, the thermal capacity of individual elements must be considered along with the sources and resistances.

Kirchhoff's mesh and nodal rules (Kirchhoff's first and second laws) apply here in thermal networks as they do in electrical networks: Heat flows (electrical currents) and temperature differentials (potential differences) may be calculated relatively easily using them. The determination of network elements is based on the system levels of electronic products (component, module, system) described in the preceding chapters.

A key benefit of the thermal network method is the high degree of abstraction it provides for a thermal design very early in the overall system design. It can be used to represent any (abstract) system with thermal resistance values. The network model is also quite intuitive, and the computational overheads involved are relatively low. This facilitates a wide range of comparisons, parameter studies, and optimizations. Disadvantages of the method include the lack of spatial resolution in the model, reduced accuracy by restriction to spatial regions, and overlapping fields interacting with each other that are not taken into account by the model.

Exemplary thermal networks are illustrated in Fig. 5.5. Such configurations of components and associated thermal network models are also described in Sect. 5.5.

Component (chip) with heat sink (hs) on printed circuit board (pcb)

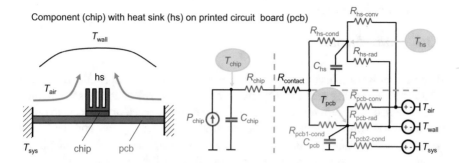

Component with heat sink in sealed enclosure (enc)

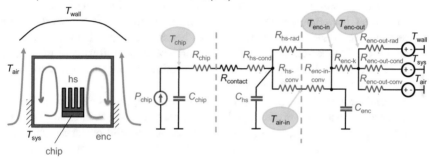

Component with heat sink in open (ventilated) enclosure

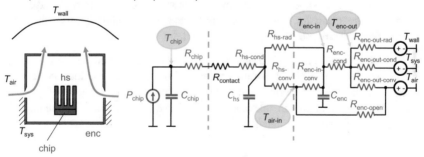

Fig. 5.5 Examples of thermal networks (subscripts cond conduction, conv convection, rad radiation; T_{chip} component junction temperature, T_{pcb} printed circuit board temperature, T_{hs} heat sink temperature, T_{air-in} inside air temperature, T_{enc-in} enclosure inside temperature, $T_{enc-out}$ enclosure outside temperature, T_{wall} temperature of irradiated surrounding walls, T_{air} outside air temperature, and T_{sys} temperature of higher-level system to mount)

5.3   Heat Transfer

5.3.1   Introduction

Heat transfer is the transportation of thermal energy due to a temperature difference. Heat is transferred between two objects or from a single object and its surroundings when a temperature differential ΔT exists between the objects, or between an object and its surroundings. The heat is transferred to the colder regions, and the heat is equalized across system boundaries. The heat transfer rate denoted by $\dot{Q}$, that is, the amount of heat energy Q transported per unit time, is the heat transfer parameter.

There are three heat transfer modes:

- In *thermal conduction* (also: *conduction heat transfer*), kinetic energy is transmitted between neighboring atoms or molecules by the movement of electrons (in conductive solids), pulse transmission, or lattice vibrations (in non-conductive solids and stationary fluids) or by the diffusion of molecules (in gases). Thermal conduction is always transmitted through a material medium and does not involve any fluid motion.
- In *thermal convection* (also: *convection heat transfer*), heat is transmitted from a solid system to a fluid stream (liquid, gas) and carried along as internal energy or enthalpy, or, vice versa, a hotter fluid flows around a solid object and "heats it up." In this case, the heat transmitting medium itself moves from one place to another.
- In *thermal radiation* (also: *radiation heat transfer*), the energy is transported by electromagnetic waves primarily in the infrared spectrum. Thermal radiation is the only heat transfer mode that can take place in a vacuum because the heat is transferred from one object to another without a medium.

Typically, there is more than one heat transfer mode in operation independently and at any given time in electronic systems. Only thermal conduction occurs in solids, whereas we find thermal conduction coupled with convection in liquids and gases. Thermal radiation is relevant when a large temperature differential exists between radiating and irradiated surfaces.

5.3.2   Conduction Heat Transfer

In thermal conduction, kinetic energy is exchanged between neighboring atoms or molecules in a material by interatomic or intermolecular forces. According to Fourier's law of thermal conduction, the amount of *heat energy Q* transmitted as per Eq. (5.6) is proportional to the temperature gradient $\Delta T/\Delta x$, the cross-sectional area of the transfer A perpendicular to the direction of heat transfer, and the

Fig. 5.6 Steady-state
conduction heat transfer
through a plane wall

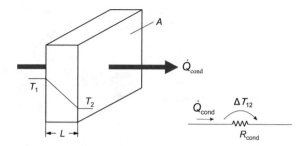

time t. The negative sign indicates that the heat flows from a higher temperature to a lower one:

$$Q = -kA\frac{\Delta T}{\Delta x}t. \tag{5.6}$$

The heat transfer rate $\dot{Q}$ is the heat energy Q transported in one time unit. The heat transfer $\dot{Q}_{cond}$ transmitted by thermal conduction through a flat wall (Fig. 5.6) calculated in Eq. (5.6) is thus:

$$\dot{Q}_{cond} = kA\frac{\Delta T_{12}}{L}, \tag{5.7}$$

where $\dot{Q}_{cond}$ denotes the heat transfer in W, k the thermal conductivity in W/(m·K), A the surface area of the wall in m², L the wall thickness in m, and ΔT_{12} the (positive) temperature differential ($T_1 - T_2$) between two sides of the wall in kelvins (K).

The proportionality constant k is called the *thermal conductivity* or *specific thermal conductance*. It is a measure of how well a material conducts heat, denoted by W/(m·K). The physical property of materials (Fig. 5.7 and Table 5.6) is composed of the thermal conductivity k_e caused by electrons and the thermal conductivity k_v arising from thermal vibrations and waves:

- Free, unbound electrons dominate in electrically conductive materials ($k_e \gg k_v$). As freely moving valence electrons transfer not only electric current but also heat energy, good electrical conductors are therefore good heat conductors (Wiedemann-Franz law).[2]
- The inequality $k_v \gg k_e$ becomes relevant in semiconductors and non-conductive solids where electrons are tightly bound, thus restricting thermal conduction to lattice vibrations only. The thermal conductivity of plastic materials, for example, is two orders of magnitude smaller than for carbon steel.

---

[2]The primary exception to this statement is diamond which has a fivefold higher thermal conductivity than copper, but a dielectric strength 10 times higher than rubber.

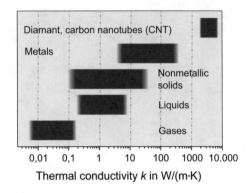

Fig. 5.7 Thermal conductivities k for different categories of materials

Table 5.6 Thermal conductivity k of selected materials at 20 °C in W/(m·K)

Metals	k	Nonmetals	k
Aluminum, pure	237	Plexiglass	0.18
AlSi1MgMn (ENAW-6082)	170	Concrete	
Aluminum oxide	10	Gravel concrete	1.3
Gold	317	Foam concrete	0.35–0.7
Copper (99.9%)	390	Soil	
Sheet copper (merchandise)	372	Coarse gravel	0.52
90% Cu, 10% Al	52	Sand (moist)	0.25–2.0
70% Cu, 30% Zn	110	Glass	1.16
Constantan (55% Cu, 45% Ni)	22	Window glass	0.8–1.1
Magnesium	156	Transparent quartz	1.45
Nickel, pure	90.7	Rubber	0.13–0.24
NiCr (80% Ni, 20% Cr)	12	Phenolic paper (Pertinax)	0.15
Silver	419	Wood (dry)	0.1–0.3
Silicon	148	Paint	0.2
Iron, pure	80.2	Leather	0.15
Steel sheet	59	Air (standard pressure)	0.0257
Low-alloy carbon steel 40Cr1Mo28 (1.7225)	43	Paper, cardboard	0.13–0.18
Austenitic CrNi carbon steel (1.4301)	15–17	Polyvinyl chloride PVC	0.12–0.25
Cast iron	58	Polyamide	0.24–0.3
Tin	67	Porcelain	0.8–1.1
SnCu (99.3% Sn, 0.7% Cu)	65	Chamotte	0.5–1.2
SnAg (96.5% Sn, 3.5% Ag)	78	Teflon	0.25
Zinc	116	Water	0.598
Bismuth	8		

– Heat energy is transported in liquids and gases by close-range vibrations, by collisions between particles, and by particle motion. Thermal conduction in these materials is much less effective than in solids. We note that the thermal conductivity of gases increases with temperature.

The thermal resistance to conduction, the *conduction thermal resistance* R_{cond}, decreases with increasing thermal conductivity (k) and larger surface area (A), and increases with growing distance (x). We can determine R_{cond} for one-dimensional thermal conduction along the x-coordinate from Fourier's law as:

$$R_{cond} = \int_{x_1}^{x_2} \frac{1}{k(x) \cdot A(x)} dx. \tag{5.8a}$$

Thus, we obtain the thermal resistance through a flat, single-layer panel of thickness L and surface area A as

$$R_{cond} = \frac{L}{k \cdot A}. \tag{5.8b}$$

Formulas for calculating conduction thermal resistances for enclosures, contacts, and insulation are given in Table 5.7. As the table shows, the overall thermal resistance of composite panels is derived from the series or parallel structure of the thermal resistances of the individual layers. If a panel consists of a number of layers, the resulting thermal resistance is the sum of the thermal resistances of the individual layers through which the heat is transported.

The thermal resistance of metallic enclosure panels in electronic systems with a maximum thickness of 5 mm is typically neglected in thermal conduction calculations. The conduction thermal resistance of an approximately 50-μm-thick paint finish is about the same as a 5-mm-thick enclosure panel made of sheet steel. (Section 5.5.6 later in this chapter contains real design examples to illustrate this.) The inside and outside finishes and the enclosure panel are treated as a series structure. The same applies to plastic coatings.

Having calculated the conduction thermal resistance, we can now calculate the temperature differential between the inside and outside surfaces of an enclosure panel by rearranging Eq. (5.7) thus:

$$\Delta T_{12} = T_1 - T_2 = \dot{Q}_{cond} \cdot R_{cond}. \tag{5.9}$$

5.3.3   Convection Heat Transfer

Heat transfer associated with material transport at the interface between a solid and a fluid, that is, between stationary and flowing media, is referred to as *convective*

Table 5.7 Application examples of conduction thermal resistances in x-direction

Description	Diagram	Conduction thermal resistance
Composite panel, layers in series		$R_{\text{cond\_tot}} = \sum_i R_i$
Composite panel, parallel layers		$\dfrac{1}{R_{\text{cond\_tot}}} = \sum_i \dfrac{1}{R_i}$
Coaxial cylinder, plated-through contact for printed circuit boards		$R_{\text{cond}} = \dfrac{L}{k \cdot \pi \cdot (r_{\text{out}}^2 - r_{\text{in}}^2)}$
Coaxial cylinder, wire sheathing (insulation)		$R_{\text{cond}} = \dfrac{\ln(r_{\text{out}} / r_{\text{in}})}{2 \cdot \pi \cdot k \cdot L}$

heat transfer, often referred to as *convection*. In this scenario, heat energy is transmitted from a solid object to an adjacent flowing fluid, such as air, and transported as intrinsic energy. The reverse scenario, that is a solid object being heated by a fluid flowing along its surface, rarely occurs in electronic systems and will hence not be further discussed.

There are two types of convection, *natural* or *free convection* and *forced convection* depending on the propagation of the fluid. Buoyancy (or lift forces) causes motion of the fluid with free convection. The lift forces arise from the density differences in the fluid due to temperature differentials. In the case of forced convection, on the other hand, external forces, such as fans, cause motion of the fluid. The convective heat transfer to the flowing liquid is thus considerably improved by this measure.

There are two types of flow, *laminar* and *turbulent*, depending on the flow stream that develops. Laminar flow contains no swirling or turbulence. The fluid flows in layers that do not mix. Turbulent flow, in contrast, is characterized by swirling and cross-flow. Laminar flows can be converted easily to turbulent flow by placing a wire mesh in the flow path. Generally speaking, turbulent flow improves heat transfer by convection.

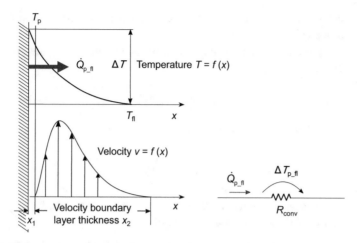

Fig. 5.8 Natural convection between a vertical panel (p) and a fluid (fl)

Heat transfer from an enclosure panel to ambient by free convection is shown schematically in Fig. 5.8. A thin, stationary layer (x_1) of the surrounding medium, e.g., air, is attached to the enclosure panel by adhesive forces; heat is transferred through this layer by thermal conduction. Heat convection in the contiguous velocity boundary layer (x_2) increases with the distance from the wall until it reaches a maximum from which it decreases again: The air is thus heated and cooled. The air flows upward (heating) or downward (cooling) due to the density differences caused by different temperatures. Based on the above classification, this is called free convection (buoyancy or downthrust forces due to density differences).

The stationary layer (x_1) is a few molecular layers wide for vertical panels or boards that are 5–50 cm thick. The temperature falls linearly in this layer, shown wider in Fig. 5.8 for the purpose of clarity.

The thickness x_2 of the velocity boundary layer adjacent to the thin, stationary layer (x_1) for vertical, heated boards at an air temperature $T_{Fl} = 20$ °C, $\Delta T \leq 60$ K, and with natural convection is

- $x_2 \approx 0.4$–0.8 cm   for board height $H = 5$ cm,
- $x_2 \approx 0.5$–1.5 cm   for board height $H = 10$ cm,
- $x_2 \approx 1.0$–2.0 cm   for board height $H = 50$ cm.

For example, the clearance s between cooling fins or parallel heated boards should be

$$s \approx 2 \cdot x_2 \tag{5.10}$$

for the best flow distribution. We recommend a clearance of at least 3 cm for printed circuit boards with heat dissipating components placed on both sides and with free convection, and we recommend a clearance of at least 1–2 cm for forced convection depending on the board height.

The convection heat transfer rate $\dot{Q}$ between panel (with surface area A) and fluid is expressed as

$$\dot{Q}_{conv} = h_c \cdot A \cdot \Delta T_{p\_fl}, \tag{5.11}$$

and the *convection thermal resistance* R_{conv} is calculated as

$$R_{conv} = \frac{1}{h_c \cdot A} \tag{5.12}$$

for what is known as Newtonian cooling. The *convection heat transfer coefficient* h_c is measured in $W/(m^2 \cdot K)$ and is a function of

- the shape, the dimensions, and the temperature of the heat dissipating surface,
- the type, temperature, and speed of the fluid, and
- the type of fluid flow (laminar or turbulent).

It is very difficult to describe analytically with differential equations the convection heat transfer, in general, and the convection heat transfer coefficient h_c, in particular, as they are impacted by many factors that are difficult to model accurately. Convective heat transfer was analyzed analytically for the first time using similitude models at the beginning of the twentieth century. Dimensionless groups are derived from the individual differential equations and have been named after well-known researchers in the field of thermodynamics. In order to evaluate the convection heat transfer coefficient h_c, the following four groups are of importance:

- The *Nusselt number Nu* is a dimensionless number, which is derived from the characteristic length L of the examined physical entity exposed to the fluid flow and the thermal conductivity k of the fluid. The Nusselt number is a measure of the ratio of the convection to conduction heat flux through a fluid layer. More precisely, it compares the intensity of a convective heat transfer process at the surface of a solid object to the theoretical, exclusive thermal conduction of a stationary fluid at this boundary. Hence, it characterizes the increase in heat transfer due to fluid motion. The Nusselt number is derived from the differential equation of heat transfer:

$$Nu = \frac{h_c \cdot L}{k} \tag{5.13a}$$

or

$$h_c = Nu \frac{k}{L}. \tag{5.13b}$$

- The *Prandtl number Pr* is the relationship between kinematic viscosity v and thermal diffusivity a:

$$Pr = \frac{v}{a} = \frac{v \cdot c_p \cdot \rho}{k}, \tag{5.14}$$

with the specific heat c_p of the fluid, its density ρ, and thermal conductivity k. The Prandtl number describes the relationship between velocity and temperature distribution; it is a measure of the relative thickness of the fluid's velocity boundary layer compared to the temperature change in the fluid. A higher number means that the velocity boundary layer is larger than the range of temperature change, i.e., heat diffuses more slowly compared to the velocity (momentum). Gases have a Prandtl number close to one.

- The *Grashof number Gr* is an expression of the ratio of buoyant forces in a fluid to the effective viscous force and is thus key for the formation of natural convection:

$$Gr = \frac{g \cdot \beta \cdot \Delta T \cdot L^3}{v^2}, \tag{5.15}$$

with gravity g, coefficient of thermal expansion of the fluid β, temperature differential between the panel and the flowing fluid ΔT, characteristic length L, and kinematic viscosity v.

- The *Reynolds number Re* is used instead of the Grashof number for the case of forced convection. It represents the ratio of inertial to viscous forces in a flowing fluid:

$$Re = \frac{v_f \cdot L}{v}, \tag{5.16}$$

with fluid flow velocity v_f, characteristic length L and kinematic viscosity v.

The ratio of the Grashof and the Reynolds numbers shows whether natural or forced convective forces are dominant. The Nusselt number, which is the key to finding the convection heat transfer coefficient h_c, can be derived from the following relationships: $Nu = c_1 \cdot (Gr \cdot Pr)^{n1}$ for   natural   convection   and $Nu = c_2 \cdot Re^{n2} \cdot Pr^m$ for forced convection. (Please refer to [2] for information on determining the coefficients c and exponents n, m.)

These relationships between the dimensionless numbers need to be determined only for iconic configurations by experimentation. They can then be applied for calculating h_c for geometrically similar configurations with any fluids by following the rules laid out in Table 5.8.

Table 5.8 Procedure for determining the convection heat transfer coefficient h_c with dimensionless numbers (see [2] for coefficients c and exponents n, m)

Step	Natural convection	Forced convection
1. Calculate the dimensionless groups	Gr, Pr	Re, Pr
2. Select suitable equation	$Nu = f\,(Gr,\,Pr)$	$Nu = f\,(Re,\,Pr)$
3. Calculate Nu	$Nu = c_1 \cdot (Gr \cdot Pr)^{n1}$	$Nu = c_2 \cdot Re^{n2} \cdot Pr^m$
4. Calculate h_c	$h_c = \frac{Nu \cdot k}{L}$	$h_c = \frac{Nu \cdot k}{L}$

Thus, we can also determine the *convection thermal resistance* R_{conv} for any surface A from the Nusselt number Nu and the thermal conductivity k of the fluid:

$$R_{conv} = \frac{L}{Nu \cdot k \cdot A} = \frac{1}{h_c \cdot A}. \tag{5.17}$$

An alternative to this laborious method of determining the heat transfer coefficient h_c is provided in Table 5.9. Here, custom dimensional equations, experimentally determined for typical electronic system configurations and assuming natural convection, are given [3]. They enable a simple and fast calculation of the convective heat transfer coefficient h_c for different geometries depending on the fluid (i.e., air or water), the fluid temperatures, and the heat dissipating surface. Applied to "typical" electronic system enclosures placed within a large room, the results are within a 10% deviation from the above-mentioned method using dimensionless numbers.

Referring to Fig. 5.9, we see that the convection heat transfer coefficient is between 5 and 10 W/(m^2·K) for free convection with air. Clearly, natural convection is often not sufficient for higher packing densities. Typically, forced convection provided by fans is used to achieve values between 10 and 120 W/(m^2·K). Even higher values can be obtained with water cooling. Figure 5.9 shows that cooling with water allows a convection heat transfer coefficient h_c that is 100–600 times higher, resulting in a 100–600 times larger heat transfer rate $\dot{Q}$ as compared to free convection with air.

5.3.4 Radiation Heat Transfer

Radiation heat transfer differs from the above-mentioned heat transfer modes, conduction and convection, since it requires no medium to occur. An object emits thermal radiation when heated. Thermal radiation takes place via electromagnetic waves between solid or liquid objects having different surface temperatures. The wavelength spectrum lies between 0.1 and 100 μm; it thus covers not only the infrared spectrum, but also the visible light and ultraviolet radiation. The spectrum of emitted radiation is continuous for solid and liquids and essentially depends only on the temperature, as shown in Fig. 5.10.

Table 5.9 Simplified calculation of the convection heat transfer coefficient h_c for the natural convection of air and water at standard pressure and with basic geometries, assuming infinite space [3]

L in m h_c in W/(m² K)	Laminar flow $\Delta T \le (0.84/L)^3$	Turbulent flow $\Delta T > (0.84/L)^3$
Vertical plate or cylinder with height L	$h_c = c_{\text{lam}}\left(\frac{\Delta T}{L}\right)^{0.25}$	$h_c = c_{\text{turb}}(\Delta T)^{0.33}$
Horizontal plate, heat dissipation from the top	$h_c = 1.3 \cdot c_{\text{lam}}\left(\frac{\Delta T}{L_{\min}}\right)^{0.25}$	$h_c = 1.3 \cdot c_{\text{turb}}(\Delta T)^{0.33}$
Horizontal plate, heat dissipation from the bottom	$h_c = 0.7 \cdot c_{\text{lam}}\left(\frac{\Delta T}{L_{\min}}\right)^{0.25}$	$h_c = 0.7 \cdot c_{\text{turb}}(\Delta T)^{0.33}$

Coefficients c_{lam} and c_{turb} for mean temperature T_{m} between air and plate surface

T_{m} (°C)	0	20	40	60	80	100
c_{lam} (air)		1.38	1.34	1.31	1.29	1.27
c_{lam} (H_2O)		105	149	178	205	227
c_{turb} (air)	1.69	1.61	1.53	1.45	1.39	1.33
c_{turb} (H_2O)	102	198	290	363	425	480

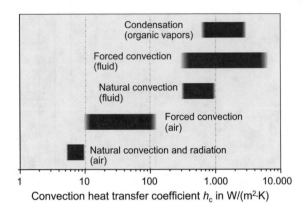

Fig. 5.9 Convection heat transfer coefficient h_c for different cooling methods

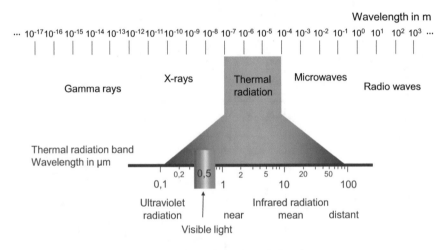

Fig. 5.10 Thermal radiation spectrum from ultraviolet to infrared

The relation between temperature and spectral distribution is described by Planck's law. The surface temperature of the sun is $T \approx 5600\,\text{K}$, and the maximum radiation emitted by the sun lies approximately in the middle of the visible electromagnetic spectrum. Window glass is largely transmissive for this wavelength range. Short-wave thermal radiation emitted by the sun generally passes through the window glass along with visible light. This radiation is partially or fully absorbed and reconverted to heat (molecular vibration) when it hits matter. The resulting temperatures are so low that the resulting radiation wavelengths cannot penetrate the window glass. This is how greenhouses are heated (the "greenhouse effect"); the passenger compartment of an automobile parked on a sunny day is another example of this type of heating.

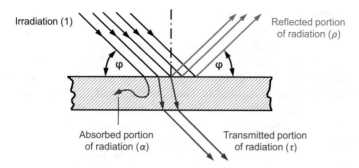

Fig. 5.11   Absorbed (α), reflected (ρ), and transmitted radiation (τ) when thermal radiation strikes an object

Thermal radiation is subject to the laws of optics, i.e., such radiation propagates linearly. This is why fins on the inside of heat sinks do not emit any heat through radiation to their surroundings: Heat is emitted only by the external enveloping surface.

When thermal radiation hits an object, it can be absorbed (proportion α), reflected (proportion ρ), or transmitted (proportion τ) (Fig. 5.11). The absorbed radiation is re-emitted as heat energy. If we set the radiation intensity striking an object to 1, we can express the sum of the absorbed radiation (α), the reflected radiation (ρ), and the transmitted radiation (τ) as follows:

$$\alpha + \rho + \tau = 1. \tag{5.18}$$

We call a body a *black body*, when all incident radiation is absorbed ($\alpha = 1$, $\rho = 0$, $\tau = 0$); a *white body*, when all radiation is reflected ($\alpha = 0$, $\rho = 1$, $\tau = 0$); and a *gray body*, when all radiation is transmitted ($\alpha = 0$, $\rho = 0$, $\tau = 1$). The terms "black," "white," and "gray" signify physical characteristics here; they do not refer to the color.

Every object can emit radiation. According to Kirchhoff's law, the absorptivity (α) and emissivity (ε) of an object are directly proportional:

$$\alpha / \varepsilon = \text{constant}, \tag{5.19a}$$

or

$$\text{absorptivity } \alpha = \text{emissivity } \varepsilon. \tag{5.19b}$$

The black body (with $\alpha = 1$) thus emits the maximum theoretical amount of radiation ($\varepsilon = 1$). A white body, which does not by definition absorb or transmit radiation, does not emit any radiation either.

Gray bodies reflect the same proportion of all wavelengths in the incident radiation, while certain wavelengths predominate with the so-called *colored bodies*. Please note that long-wave thermal radiation is colorless. Colored surfaces are generally thermally gray to black. Incident thermal radiation is largely absorbed, and the objects emit thermal radiation commensurate with their temperature and emissivity. Short-wave, visible thermal radiation, which is reflected to a greater degree from white surfaces than from colored ones, is found only in sunshine.

According to Eqs. (5.18), (5.19a), and (5.19b), opaque surfaces ($\tau = 0$) with high reflection always have low absorption and emission and vice versa. While bright metals have low absorption and emissivity, nonmetals and metal oxides have relatively high absorption and emissivity. Metals can be made good radiators (take, e.g., cooling surfaces and heat sinks for semiconductors) if they are finished, oxidized, and anodized, as the surface composition is the key factor for thermal radiation.

The (relative) *emissivity* ε, which is the ratio of the emissivity of any object to the emissivity of a black body (ideal heat radiator) and where $0 \leq \varepsilon \leq 1$, should be used for practical applications. Thus, the emissivity $\varepsilon = 1$ for an ideal heat radiator; this is the theoretical maximum value for heat emission. And, conversely, when the emissivity $\varepsilon = 0$, the object has no emissivity.

It is assumed that temperature does not impact the emissivity for electronic systems operating under standard temperature conditions. Please note that for finished materials, the surface finish is the determining factor for ε and not the actual panel material. Specifically, the emissivity depends only on the surface of the finish for finish thicknesses >30 µm.

Table 5.10 and Fig. 5.12 show the emissivity ε for materials commonly found in electronic systems, where the values are based on the standard temperature range (250 K < T < 470 K) and on a vertical angle of emission with respect to the surface. As the emissivity ε is similar to the absorptivity α, bright metals with their low absorptivities have low ε values, whereas organic materials and oxides have large ε values. Metallic paints have low emissivities in contrast to other paints.

Objects *exchange* radiation with their surroundings because all objects at temperatures above zero kelvin emit radiation. Radiating objects include components, modules, panels, and the like. Air emits almost no thermal radiation, as it does not absorb any radiation. An object with a higher surface temperature T_1 transfers more radiation to the surface of another object at a lower temperature T_2 than vice versa. The resulting heat transfer rate or the transmitted radiation is proportional to the difference between the fourth powers of the two temperatures in degrees kelvin:

$$\dot{Q} \sim T_1^4 - T_2^4. \tag{5.20}$$

There is a radiation exchange between a surface area A_1 which is at a higher temperature T_1 and a parallel surface area A_2 at a lower temperature T_2 (Fig. 5.13). According to the Stefan–Boltzmann law, the *heat transfer rate* between these two surfaces is:

$$\dot{Q}_{rad} = h_r \cdot A_1 \cdot \Delta T_{12}. \tag{5.21}$$

Table 5.10 Emissivity ε for standard radiation (mean values)

Metals	ε	Nonmetals	ε
Aluminum, rolled, plain	0.04	Ice, water	0.95
Aluminum, oxidized	0.25	Oak, smooth	0.9
Aluminum, anodized, 30 μm layer	0.65	Enamel, white	0.9
Chrome, bright	0.08	Glass	0.94
Cast iron, raw	0.9	Rubber, soft	0.9
Cast iron, treated	0.7	Masonry	0.91
Copper, bright	0.03	Paper	0.92
Copper, slightly/greatly oxidized	0.25/0.76	Porcelain, glazed	0.93
Brass, bright	0.05	Teflon	0.85
Brass, mat	0.22		
Nickel, bright	0.07	Finishes	ε
Nickel, oxidized	0.4	Aluminum paint	0.3
Silver, bright	0.02	Enamel	0.9
Carbon steel, rolled	0.6	Hammer finish	0.35
Carbon steel, slightly rusty	0.7	Paint, black, high-gloss	0.89
Carbon steel, very rusted	0.85	Paint, black, mat	0.96
Carbon steel, brightly sanded	0.24	Paint, white, mat	0.92
Carbon steel, brightly etched	0.13	Red lead	0.92
Steel sheet, wrought	0.6	Oil paint	0.9
Steel sheet, galvanized	0.27	Special aluminum paint	0.2
Steel sheet, nickel-plated, unpolished	0.11		
Tin, bright	0.06		
Zinc, bright	0.05		
Zinc, oxidized	0.11		
Zinc, raw	0.25		

Fig. 5.12 Emissivities ε for different surfaces

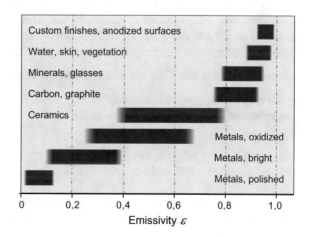

Fig. 5.13 Thermal radiation
between two vertical surfaces
A_1 and A_2

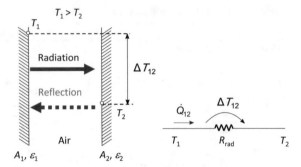

The constant h_r is the radiation heat transfer coefficient (see below), A_1 is the radiation surface area, and ΔT_{12} is the temperature differential between radiating and absorbing surfaces ($T_1 > T_2$).

We can express the *radiation thermal resistance* as follows:

$$R_{rad} = \frac{1}{h_r \cdot A_1},\qquad(5.22)$$

similar to convection, where A_1 is the heat dissipating surface area here, as well.

The *radiation heat transfer coefficient* h_r is expressed as follows:

$$h_r = \varepsilon_{res} \cdot \sigma \cdot \frac{T_1^4 - T_2^4}{\Delta T_{12}},\qquad(5.23a)$$

in units W/(m²·K), with the Boltzmann constant $\sigma = 5.67 \cdot 10^{-8}$ W/(m²·K⁴) and the resulting emissivity ε_{res} for the two surfaces. The temperature of the heat dissipating surface is T_1; T_2 is the temperature of the irradiated surface; and ΔT_{12} the temperature differential between the two surfaces. All temperatures are in kelvin (K).

The size of the *resulting* emissivity ε_{res} for both surfaces depends on the shape, size, position, and orientation, clearance and surface area of the two objects involved in the radiation exchange. Four sample calculations (a) to (d) are given below for geometries that frequently occur in practice.

General scenario (a): One surface completely surrounded by another surface

This general scenario is depicted in Fig. 5.14. The emissivity is approximated regardless of the shape and distance as follows:

$$\varepsilon_{res} = \frac{1}{\frac{1}{\varepsilon_1} + \frac{A_1}{A_2}\left(\frac{1}{\varepsilon_2} - 1\right)},\qquad(5.24)$$

where A_1 denotes the surface area of the smaller, enclosed object in m², A_2 denotes the surrounding surface area in m², and ε_1 and ε_2 are the emissivities of surfaces A_1 and A_2, respectively.

Fig. 5.14 General
scenario (a) where the heat
dissipating object 1 is fully
enveloped by surface 2

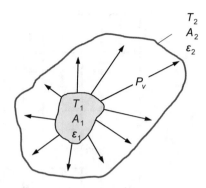

Special case (b): Two fully enveloping surfaces with $A_2 \gg A_1$

This is a special case of (a) that occurs often in electronic systems, when a small, hot component (A_1) is located in a much larger enclosure (A_2) or an emitting electronic system is installed in a room. We can simplify Eq. (5.24) for $A_2 \gg A_1$ to:

$$\varepsilon_{res} \approx \varepsilon_1. \tag{5.25a}$$

For this scenario, which is commonly encountered in practice, we can quickly calculate the radiation heat transfer coefficient h_r using the chart in Fig. 5.15 [instead of using Eq. (5.23a)]. First, we read off the radiation heat transfer coefficient h_r^* of the black body as a function of the surface temperature of the radiating surface A_1 and the surface temperature of the enveloping surface A_2. We then calculate the radiation heat transfer coefficient h_r by multiplying h_r^* with the emissivity ε_1 of the radiating surface as per Eq. (5.23b):

$$h_r = \varepsilon_1 \cdot h_r^*. \tag{5.23b}$$

Special case (c): Two parallel, flat surfaces with $A_1 \approx A_2$

This scenario serves to approximately calculate the heat transfer between two printed circuit boards or one printed circuit board and an enclosure panel (Fig. 5.16 left). The *effective* surface area $A_1 \approx A_2$ is the same for both parallel, flat surfaces separated by a small distance, which allows us to simplify the calculation of the resulting emissivity as per Eq. (5.24) as:

$$\varepsilon_{res} = \frac{1}{\frac{1}{\varepsilon_1} + \frac{1}{\varepsilon_2} - 1}. \tag{5.25b}$$

Special case (d): Partition panel effect

A partition panel is introduced in this example to protect modules from radiation (Fig. 5.16 right). We calculate the resulting emissivity ε_{res} as:

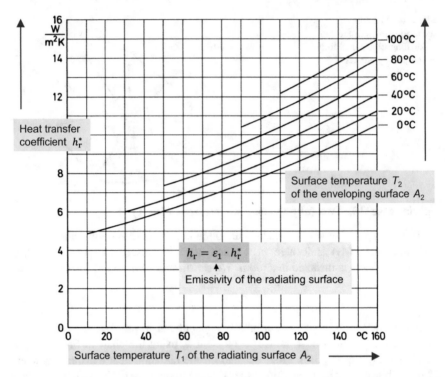

Fig. 5.15 Obtaining the radiation heat transfer coefficient h_r based on the surface temperatures for fully enveloping surfaces where $A_2 \gg A_1$. First, determine h_r^* on the vertical axis of the graph above. Then, multiply this value with the emissivity ε_1 of the radiating surface A_1 as per Eq. (5.23b)

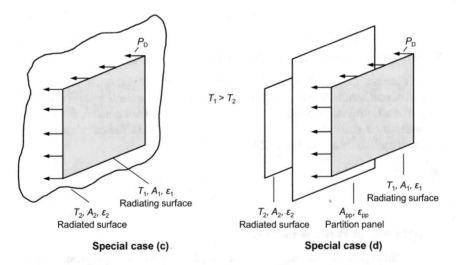

Fig. 5.16 Special case (c) comprises two parallel, flat surfaces, common for thermal radiation from a printed circuit board. The use of a partition panel (special case (d)) is shown on the *right*

$$\varepsilon_{res} \approx \frac{\varepsilon}{2},$$

(5.25c)

with $\varepsilon_1 = \varepsilon_2 = \varepsilon_{pp} = \varepsilon$ and $A_{pp} \gg A_1, A_2$ and $A_1 \approx A_2$. Hence, the heat transferred from A_1 to A_2 is approximately halved by using a partition panel.

More complex geometries composed of irregular surfaces or surfaces arranged at an angle to one another that are not covered by scenarios (a) to (d) are typically divided into finite surface elements, and the amount of radiation exchanged is determined for each of the finite elements. *View factors* can then be calculated, which express the proportion of the emitted radiation striking the absorbing surface, and are based on the distance, the surface area, and the deviation from the vertical radiation angle [1].

The examples given in Sect. 5.3.4 show that a high emissivity ε_{res} is required for a large radiation heat transfer (see Eqs. (5.21) and (5.23a)). Additionally, the heat dissipating surface should be large as per Eqs. (5.21) and (5.22). Furthermore, the effects of thermal radiation increase with a higher temperature differential between the radiator and the radiated surface, as illustrated by Eq. (5.21). Please refer to Examples 5.2 and 5.3 in Sect. 5.5.5 for more practical calculations.

5.4 Methods to Increase Heat Transfer

Heat generated in electronic components must be dissipated with minimum resistance to prevent exceeding temperature limits, such as maximum junction temperatures in semiconductor devices. The heat transfer mechanisms outlined in Sect. 5.3 should be used selectively and judiciously for this purpose. If required, an array of different devices is available to provide more powerful heat transfers. These devices will be described in the following sections.

5.4.1 Heat Sinks

A heat sink is a passive heat exchanger that transfers the heat generated by a component to a fluid medium, often air, where it is dissipated. The heat transfer is intensified in heat sinks through improved convection and radiation due to a surface area that is larger than that of the component. The surface areas of heat sinks are enlarged by either fins or pins. Heat sinks are deployed for free and forced convection. Their purpose is as follows:

- to transport heat by thermal conduction from heat generating components and
- then to dissipate it to the surroundings by convection and thermal radiation.

Obviously, a heat sink should be constructed with good heat-conducting materials and have a large surface area. Heat sinks are typically made of aluminum or aluminum alloys. Copper is also commonly used due to its good thermal conductivity; it is however heavier, more expensive, and difficult to machine. Different heat sink designs are pictured in Fig. 5.17.

The principle heat sink dimensions are shown in Fig. 5.18.

The fin pitch s on a heat sink is selected so that the dissipated heat transfer per heat sink volume is maximized. The fin pitch s should be chosen so that the velocity boundary layers of opposite fins partially overlap, thus yielding:

$$x_2 \leq s \leq 2x_2, \tag{5.26}$$

similar to Eq. (5.10) and with the velocity boundary layer thickness x_2 of natural convection (see Fig. 5.8).

Fig. 5.17 Different heat sink designs: finned heat sinks (*left* and *right*), lamella fin heat sinks (*center*), and pin heat sinks (at *back right*). While finned heat sinks are applied for natural convection, lamella fin heat sinks are used for forced convection due to their smaller fin pitches

Fig. 5.18 Typical dimensions of heat sinks: width W, length L, fin height h, distance s between fins, and base width d

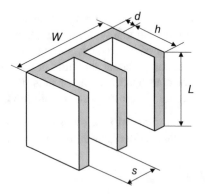

Heat sinks designed for natural convection have much larger fin pitches, due to their relatively wide velocity boundary layer, than those designed for forced convection. Lamella fin heat sinks made of carbon steel lamellas are typically used as heat sinks for forced convection because of their smaller fin pitches (see Fig. 5.17).

The *optimal fin pitch* can be determined with custom dimensional equations for free air convection with:

$$s_{opt} = 1.3 \cdot \sqrt[4]{\frac{h \,/\, cm}{(T_{fin} - T_{air}) \,/\, K}} \cdot cm \tag{5.27a}$$

and for forced convection with:

$$s_{opt} = 0.4 \cdot \sqrt[2]{\frac{h \,/\, cm}{v \,/\, \frac{m}{s}}} \cdot cm, \tag{5.27b}$$

where h is the fin height, T_{fin} the fin temperature, T_{air} the fluid temperature (air), and v its flow velocity [6].

Please refer to the manufacturer's specifications for the *thermal resistance* R_{th} of heat sinks (Fig. 5.19). The thermal resistance R_{th} is generally specified for heat sinks as a function of the length L for natural convection (see Fig. 5.18) and as a function of the flow velocity v for forced convection.

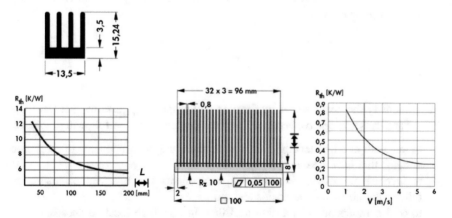

Fig. 5.19 Catalog specifications are based on the length L (*left*) for the thermal resistance R_{th} for finned heat sinks for natural convection, and on the flow velocity v (on the *right*) for lamella fin heat sinks for forced convection [4]

5.4.2   Thermal Interface Materials

A thermal contact resistance exists when two solids are in contact, e.g., between a heat-producing electronic component and a heat sink. This is due mainly to the imperfect connection between the two objects related to surface roughness, air infiltration, and the like.

Thermal interface materials are used to improve the heat transfer at these contact surfaces. Thermal glues and greases are typically applied for electronic components and modules. Thermal gap fillers and thermal pads with and without adhesive properties are available for use as well.

Thermal glues are adhesive sealant pastes made of epoxy resin or silicone with added ceramic (e.g., ZnO, Al_2O_3, SiO_2) or metallic (e.g., Al, Ag, Cu) filling materials for making a permanent contact that will have a very small thermal resistance. They are typically used in electronic systems to glue heat sinks to electronic components. Thermal adhesives that are 100 μm thick have surface thermal resistances of approximately 1 K·cm²/W for a thermal conductivity of approximately 1 W/(m·K).

Thermal greases also improve the contact between solids with their good wetting properties; they have even lower thermal resistances than thermal glues. Thermal greases are used with detachable connections, e.g., fitting heat sinks to electronic components, that require a separate mechanical fixation mechanism. Their filling materials are similar to those in thermal glues; they can be electrically conductive or non-conductive. Customary thermal greases offer surface thermal resistances down to 0.03 K·cm²/W for a thermal conductivity of up to 10 W/(m·K) for a 30 μm thickness.

5.4.3   Fans

Fans intensify the heat transfer by forming and propagating airflows that transport heat. Fans cause *forced convection*. They are deployed for the following reasons:

- to maintain at a low level the ambient temperature of components in electronic systems that have openings in their enclosures;
- to increase the convection between components and a sealed enclosure by mixing the air inside the enclosure; and
- to cool components with a high heat flux through high local air speeds.

Figure 5.20 pictures typical fans used in electronic systems. *Axial-flow fans* are the most commonly used (Fig. 5.20 left). The axis of rotation of the impeller is in the axial direction and is thus parallel to the airstream. Its efficiency depends on the speed. *Axial-flow fans*, also called *axial fans*, produce low outlet pressure for a medium volumetric flow rate. In a *centrifugal fan*, also called *radial fan* or *blower*, air enters the impeller parallel to its axis of rotation and leaves it in the radial

Fig. 5.20 Axial-flow fan (*left*), centrifugal fan (*center*), and tangential fan (on the *right*)

direction (Fig. 5.20 center). Centrifugal fans are used in applications where higher pressures than those produced by axial fans are needed for the same volumetric flow rate. *Tangential fans*, also called *cross-flow fans*, have a wide, cylindrical impeller with many small blades (Fig. 5.20 on the right). Air is drawn tangentially into the impeller over a large area approximately more than half the impeller surface, and it is routed through the wheel and exhausted again tangentially. Tangential fans can produce large volumetric flow rates at medium pressures evenly over a wide exhaust area. They operate at very low noise levels due to their low rotational speed.

Fans can be fitted in air inlets or air outlets in an enclosure, or in a bulkhead partition in the system itself. Multiple fans arranged in parallel increase the volumetric flow rate, and multiple fans in series increase the outlet pressure (see Sect. 5.5.10).

Every fan produces a *static pressure* Δp (also called *pressure rise, outlet pressure,* or *dynamic head*), necessary to overcome the resistance to airflow (air friction) and to cause the air moving through the system (air motion). The volumetric flow rate $\dot{V}$ is the volume of air, transported per time unit through a predefined cross-sectional area of the flow channel. A characteristic *fan curve* is produced (Fig. 5.21) by plotting the static pressure Δp produced by the fan versus the volumetric flow rate $\dot{V}$ when going from an uninterrupted aerodynamic flow (free flow) all the way to an aerodynamic stall (shut off, see Sect. 5.5.10 for more details.). Fan-rating curves are provided by manufacturers.

In order to evaluate the fan's performance in a particular electronic system, the graph (see Fig. 5.21) also contains a *system pressure curve* (also: *system impedance curve* or *head loss*) for the system to be ventilated. This curve plots the system-specific relationship between an achievable volumetric flow rate $\dot{V}$ and the static pressure Δp. The trajectory of this characteristic curve is different for every system depending on the air resistance of the flow channel in the system; the "flatter" the curve, the more "open" (unrestricted) the electronic system is to the airflow (see Fig. 5.21).

Any given fan can only deliver one flow at one pressure in a particular system. This *operating point* of a fan is determined by the intersection of the fan curve with the system pressure curve (see Fig. 5.21). Fans should be operated in the lower

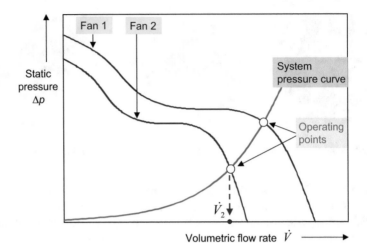

Fig. 5.21 Interaction of two fan curves and a system pressure curve of a given electronic system. Since an operating point in the lower range of the fan curve (i.e., less pressure) minimizes fan acoustic noise, fan 2 should be chosen, provided its volumetric flow rate $\dot{V}_2$ is sufficient for cooling

third region of the pressure range on their characteristic curve to reduce acoustic noise from them. The fan operating point is often used to determine which fan should be used for a given application.

We shall discuss practical considerations regarding fans, such as determining the system curve and fan selection, in Sect. 5.5.10.

5.4.4   Heat Pipes

A heat pipe combines the principles of both thermal conductivity and phase transition to efficiently manage the transfer of heat over a certain distance. Heat is transmitted through a vaporous medium; it is thus transported by mass transfer. Heat pipes are used in electronic systems to transfer heat from inaccessible components or places with high heat fluxes to places where heat sinks can easily be installed.

A heat pipe consists of a sealed slender pipe, containing a medium (e.g., methanol) that is easily vaporized (Fig. 5.22). The heat pipe absorbs heat at one end, the *evaporator section*; the medium vaporizes in this section. This vapor condenses at the other cold end of the tube, the *condenser section*. The heat released by the condensing process is dissipated at the end of the pipe and released to ambient by convection and radiation. Attached heat sinks are typically used for this purpose.

The vapor flows through the *transport section* between the two pipe ends due to the pressure difference, which is based on the temperature differential between the

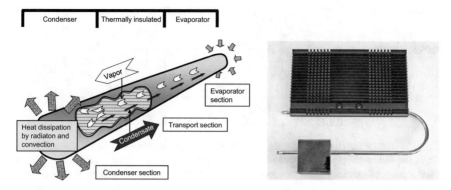

Fig. 5.22 Mode of operation of a heat pipe (*left*). A graphics processor being cooled by a separate heat sink is shown on the *right*. The bent heat pipe transports heat from the processor (*below*) to the large-surface-area heat sink (*above*). A copper bar of the same diameter and length would have approximately 250 times less thermal conductivity

two ends. The vaporization process at one end and the condensation process at the other end create a pressure gradient, which causes the hot vapor to "flow" to the colder condenser section.

The condensate flows back to the evaporator section in liquid form, either under the influence of gravity (this process is called thermosyphoning) or in capillary tube-like structures (capillary heat pipe) as a result of capillary action.

Working fluids are chosen to match the temperatures at which the heat pipe should operate, as the heat pipe must contain a saturated liquid and its vapor (gas phase) inside this temperature range. Most heat pipes for room temperature applications use ammonia, methanol, ethanol, or water.

Heat pipes are increasingly installed in electronic systems. They are used, for example, in notebooks to dissipate heat from powerful processors: The heat is transported via a heat pipe to the outside surface of the notebook. Copper envelopes with water as working fluid are commonly applied here as they cover a temperature range of 20–150 °C. Electronic modules that are used in space exploration are another critical application: Heat pipes can be used to cool and equalize the temperatures between sun-facing and sun-averted sides of space probes where the temperature gradient can be up to 130 K.

5.4.5   Peltier Elements

Thermoelectric devices directly convert temperature differences to electric voltage and vice versa. The *Seebeck effect* is the conversion of heat into electricity at the junction of different types of wire. Here, voltage is created when there is a temperature differential. With the opposite *Peltier effect*, a temperature differential is

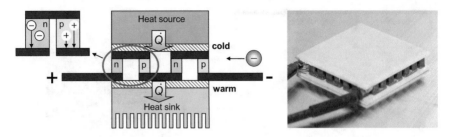

Fig. 5.23 Schematic (*left*) and practical implementation (on the *right*) of a Peltier element. In this single-stage example, *p*- and *n*-doped semiconductor pellets are connected electrically in series and when a current flows they generate a heat transfer $\dot{Q}$ from the *top* (cooled) to the *bottom* (heated) ceramic plate

produced in the conductor by an external current. This thermoelectric effect is named after French physicist *Jean Charles Athanase Peltier* who discovered it in 1834.

A *Peltier element* is a thermoelectric device that, based on the Peltier effect, produces a heat transfer and thus a temperature differential when a current passes through it. Peltier elements are commonly used as cooling devices in electronic systems.

Electrons and holes transport charge and energy. When a voltage is applied to the *p*-type region of a semiconductor as shown in Fig. 5.23 (upper left), "the holes migrate" and thus their energy from top to bottom ("hole migration" as opposed to electron migration). At the same time, the electrons and their energy move from top to bottom as well in the *n*-type region depicted in the figure. The upper electrode loses energy in both scenarios and thus is cooled. This energy is transferred to the lower electrode in the form of heat.

A Peltier element consists of coupled *p*- and *n*-doped semiconductors. They are connected thermally in parallel and electrically in series and are sandwiched between two thermally conducting and electrically insulating plates as shown in Fig. 5.23. In this configuration, the Peltier element absorbs heat at one side (here: top) and dissipates heat at the other side (here: bottom). Heat is thus actively transferred based on the Peltier effect. This produces a large temperature differential across the Peltier element depending on the operating conditions.

Industrial Peltier elements are often comprised of two metalized ceramic plates, with *p*- and *n*-doped modules, or pellets, placed between them. A heat sink is usually attached to the warm side of the element. Single-stage Peltier elements reach temperature differentials of up to 75 K, and multi-stage elements reach higher values (e.g., four-stage elements up to 127 K). We can increase the temperature differentials by connecting semiconductor modules thermally in series and increase cooling by connecting the modules in parallel.

Key benefits of Peltier elements are their small size, zero maintenance, and lack of noise. Their low efficiency is a drawback, which means high energy consumption for a relatively low cooling effect.

These elements should be deployed where low cooling effects are needed regardless of efficiency and the use of other heat dissipating elements is impractical due to space considerations. Sensor cooling, for example, CCD chips in digital cameras, is a typical application in electronic systems.

5.5   Application Examples in Electronic Systems

Every electronic component has a maximum approved operating temperature; for example, semiconductor devices are labeled with a maximum junction temperature. A key objective of thermal management is to hold the actual operating temperature below this maximum junction temperature specified by the semiconductor manufacturer. In order to achieve this, the engineer is faced with the questions as to (1) what is the expected temperature of the components for a predefined power dissipation (consumption) or (2) what is the approved power dissipation for a specified temperature limit.

The more complex the design, the more difficult it is to calculate the thermal parameters. Nevertheless, we can design and implement an effective heat dissipation scheme by applying the basic calculation methods and heat transfer mechanisms introduced in the preceding sections. Prior to this, however, every effort should be made to reduce the amount of power and, thus power dissipation, of the electronic system itself.

5.5.1   Component Temperatures

The *maximum operating temperature of components* should not be exceeded for operational and reliability reasons. These temperatures in the case of semiconductor devices are the maximum junction temperatures T_j that are between 60 and 100 °C for germanium components and between 125 and 200 °C for silicon components. Thus, the approved overtemperatures ΔT_j for these components are between 40 and 80 K (germanium) and 105 and 180 K (silicon) at a room temperature of 20 °C. If the heat losses from these components themselves and any external heating will cause component temperatures to rise beyond the above temperature limits, measures must be put in place to transfer the heat from the component to the surroundings, in order to reduce the component temperatures. To this end, the thermal resistance to the dissipation heat transfer from the component to the surroundings should be reduced. One option here is to use heat sinks to artificially increase the component surface area by *spreading the heat*. Let us now investigate how to select an appropriate heat sink to achieve this goal.

Figure 5.24 depicts the thermal resistance network for this problem. It shows that by reducing the temperature drop across the contact material and heat sink ($\Delta T_{contact\text{-}hs}$), the barrier junction overtemperature ΔT_j in components, e.g., chips, is

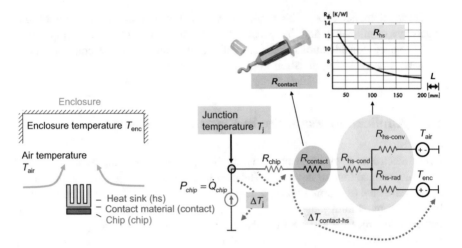

Fig. 5.24 Equivalent thermal resistance network for determining the junction temperature T_j of a component (chip). A reduction in the thermal resistances $R_{contact}$ of the contact material and the heat sink R_{hs} cause a smaller temperature drop ΔT_j across the chip, which lowers its junction temperature T_j as well

reduced as well. This is equivalent to using a heat sink with a low thermal resistance R_{hs} and thus better heat distribution and dissipation. The barrier junction overtemperature of the chip is calculated as:

$$\Delta T_j = \dot{Q} \cdot R_{th} = \dot{Q} \cdot \left(R_{chip} + R_{contact} + R_{hs}\right), \qquad (5.28)$$

with the heat transfer $\dot{Q}$ resulting from the chip's power dissipation, the predefined thermal inner resistance R_{chip} of the chip, and the thermal resistances of the contact material $R_{contact}$ and heat sink R_{hs}.

We can thus calculate and impact the component overtemperature ΔT_j by changing the thermal resistances of the contact material $\left(R_{contact}\right)$ and the heat sink $\left(R_{hs}\right)$. Obviously, the lower their thermal resistances, the lower the overtemperature of the component. By using these methods, we can also establish whether a heat sink is sufficient to remove the heat or if other elements are needed in order to keep the operating temperature of the component below its maximum value.

Considering that components are usually placed within enclosures, their thermal resistances must be included in real application settings. Simply put, if we know the enclosure's thermal resistance, the thermal resistance network in Fig. 5.24 can be extended to include the entire electronic system and ambient. This allows us to determine the component's application temperature and adjust it within the entire electronic system's configuration. With this as motivation, let us now investigate the thermal characteristics of different enclosure options.

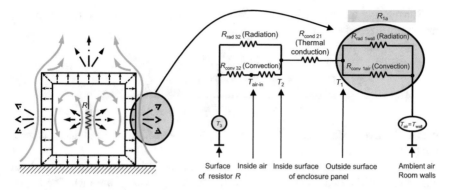

Fig. 5.25 Equivalent thermal resistance network for determining the temperature of an enclosure (see also Figs. 5.1 and 5.4). Equalizing the temperature of the ambient air with the temperature of the room walls ($T_{air} = T_{wall}$) allows us to model the thermal resistance R_{1a} between the outside of the enclosure and ambient by placing in parallel the thermal resistances for convection and radiation from the enclosure surface

5.5.2 Outside and Inside Surface Temperatures of an Enclosure

Every calculation of the temperature distribution in electronic systems should be based on the ambient temperature (which is assumed to be constant) of the room where the system is located. The *outside surface temperature of the enclosure* is independent of the thermal resistances and temperatures inside the enclosure. It is a function only of the heat to be transported and the thermal resistances between enclosure surface and surroundings.

The resulting thermal resistance R_{1a} between enclosure and ambient is based on convection and thermal radiation. As shown in Fig. 5.25, the panel outer surface of the enclosure interacts with the ambient air (T_{air}) via convection and with the surrounding building walls (T_{wall}) via radiation.

The approximation $T_{air} = T_{wall}$ (i.e., the ambient air temperature equals the temperature of the surrounding walls) can usually be used in calculations in most analyses. The thermal resistances of the convection $R_{conv\ 1air}$ and radiation $R_{rad\ 1wall}$ are thus in parallel, and the resulting thermal resistance R_{1a} between the outside surface of the enclosure and ambient can be expressed as:

$$R_{1a} = \frac{R_{conv\ 1air} \cdot R_{rad\ 1wall}}{R_{conv\ 1air} + R_{rad\ 1wall}}. \tag{5.29a}$$

Substituting the expressions for thermal resistance for convection (5.12) and for radiation (5.22) gives us:

$$R_{1a} = \frac{\dfrac{1}{h_c \cdot A_{conv}} \cdot \dfrac{1}{h_r \cdot A_{rad}}}{\dfrac{1}{h_c \cdot A_{conv}} + \dfrac{1}{h_r \cdot A_{rad}}} = \frac{1}{(h_c + h_r) \cdot A}, \tag{5.29b}$$

if the effective surface areas A_{conv} and A_{rad} for convection and radiation, respectively, are the same.

We can add together the heat transfer coefficients for convection h_c and for radiation h_r if the two conditions $T_{air} = T_{wall}$ and $A_{conv} = A_{rad}$ are met. The higher resultant heat transfer coefficient $h_{tot} = h_c + h_r$ reduces the thermal resistance R_{1a} between enclosure and ambient. As a result, the overtemperature of the enclosure surface that can be calculated as:

$$\Delta T = \dot{Q} \cdot R_{1a} \tag{5.30}$$

also falls as well.

Assuming uniform dissipation of the heat loss via the surface of a free-standing enclosure (constant heat flux), we can calculate the mean surface temperature of a sealed enclosure using this technique.

Please note that the heat transfer coefficients depend on the temperature (see Table 5.9 and Fig. 5.15). Higher temperature differentials between surface (T_1) and surroundings (T_{air}, T_{wall}) improve the heat transfer with free convection and radiation.

Having calculated the thermal resistance of the enclosure panel as per Eq. (5.8b), we can calculate the *temperature of the inside surface of the enclosure* (T_2) with Eq. (5.9). We assume a constant heat flux here as well. The remaining temperatures inside the enclosure rise steadily: starting at the inside surface temperature of the enclosure, followed by the inside air temperature (see Sect. 5.5.7), and ending at the component temperatures (see Sect. 5.5.1).

If we were to connect the surface of the resistor in Fig. 5.25 directly with the enclosure via a good heat conductor, there would be a significant reduction in the surface temperature of the resistor (T_3). This reduction in the component temperature would, nevertheless, have no impact on the external surface temperature of the enclosure (T_1), assuming a constant heat flux across the enclosure.

It is difficult to calculate the *air temperature inside enclosures* at the design stage if the convection is disrupted by components and modules, such as printed circuit boards (PCBs). Some vendors provide data on the relation between power input (P_D) and mean air temperature inside enclosures for standard modules and enclosure configurations. The combined heat transfer coefficient $h_{tot} = h_c + h_r \approx 10 \ W/(m^2 \cdot K)$ can be used for ballpark estimates. If the air temperature inside an enclosure is known, we can calculate the mean surface temperature of modules, such as PCBs, if we assume they are heated, vertical plates.

It should also be noted that vendors specify the combined thermal resistance for power semiconductors and heat sinks.

Finally, the effect of sunlight needs to be considered for systems installed outdoors. Here, the extreme conditions to which the enclosure will be exposed must be taken into account. Reflections of solar energy from surrounding surfaces can impact the radiation exposure as well. In many cases, only standardized test procedures deliver reliable values of the temperature rise inside the enclosure due to exposure to solar radiation.

5.5.3 Choosing Open or Sealed Enclosures

Many attempts have been made, including the use of empirical "rules of thumb", curves, and tables, to overcome the challenges encountered when trying to calculate temperature distributions in electronic systems. A disadvantage of these approximations is that they apply accurately only to the special cases for which they have been drafted. You should not use these aids without knowledge of the conditions and criteria under which they were drawn up in the first place.

The first thing to appreciate is that there are open (perforated) and sealed (closed) enclosures (Fig. 5.26). Air generally enters through the bottom of an enclosure in open systems, removes the heat from the module surfaces (e.g., printed circuit boards) by convection, and exits at the top. Sealed systems are needed where this mode of heat dissipation is not approved due to other system requirements, such as shock protection, protection against ingress of foreign bodies or water, and electromagnetic compatibility.

Table 5.11 provides useful insights into internal and external heat dissipation from electronic systems as related to the approved heat flux density $\dot{q}$ to be dissipated.

Table 5.11 is based on the approved standard ambient temperatures for semiconductors. The ranges for heat flux densities reflect the different clearances and

Fig. 5.26 Heat dissipation by convection in open (*left*) and sealed enclosures (on the *right*)

Table 5.11 Approved heat flux densities $\dot{q}$ of systems (power in watts per $10 \times 10 \times 10$ cm "interior space") depending on the internal and external heat dissipation from open and sealed enclosures [5]

Enclosure	Internal heat dissipation	External heat dissipation	
		Natural convection (W/dm^3)	Forced convection (W/dm^3)
Open		30–60	300–600
Sealed	Natural convection	5–15	10–60
	Forced convection	10–50	30–180
	Thermal conduction	20–90	120–240

installation positions of printed circuit boards, as well as different air speeds and temperatures with forced cooling. The external heat dissipation is assumed to be primarily convection.

To illustrate the use of heat flux densities such as shown in Table 5.11, if we consider a power supply module with dimensions $15 \times 20 \times 25$ cm = 7500 cm^3 = 7.5 dm^3 with a power consumption (and thus assumed power dissipation) of $P_D = 150$ W, then the resulting heat flux density is 20 W/dm^3. There are a number of techniques available for the heat dissipation in this example according to Table 5.11:

- Open enclosure with natural convection (i.e., passive convection for external heat dissipation).
- Closed enclosure with natural convection inside (i.e., passive convection for internal heat dissipation) and forced convection outside. We assume in this case that natural convection can form in the small enclosure.
- Closed enclosure with forced convection inside and natural convection outside. This approach is impractical due to the small size of the enclosure.
- Closed enclosure with thermal conduction for heat transfer from the interior heat sources to the enclosure and natural convection outside.

The quoted values apply only when all measures are in place to maximize the effect of the chosen technique. Table 5.11 is essentially a good guide for selecting suitable techniques for heat dissipation from open and closed enclosures. The values in the table are not suitable for very large enclosures as the surface/volume ratio is degraded with increasing system size.

Another rule of thumb states that for each watt of power, an enclosure surface area of 10×10 cm is required if we rely on natural convection outside of a *sealed* enclosure. Specifically, heat fluxes of up to 100 W/m^2 can be dissipated through a sealed, bare metallic enclosure (mostly convection) for a temperature differential of 20 K, and up to 200 W/m^2 through painted surfaces (convection and radiation, all these values are approximations). Higher values indicate that open enclosures might be required.

After providing the various options for dissipating heat from system enclosures, we now investigate the thermal characteristics of open and sealed enclosures in more detail.

5.5.4 Heat Dissipation from Open Enclosures

Table 5.11 shows the benefits of open enclosures: a large amount of heat can be dissipated with very little work (this is due to the increased level of natural convection inside the system). Intuitively, this convection depends on the number of vents in the enclosure and their arrangement, and the resulting ventilation channels in the system. However, it is very difficult to calculate the magnitude of these convection characteristics.

Based on the experimental results obtained in [6], we introduce a *perforation coefficient* ψ that allows us to efficiently estimate the increase in convection, and thus the ventilation efficiency, depending on the number and size of vents (openings) in the enclosure. Specifically, the perforation coefficient defines the percentage of the enclosure surface that is effectively perforated, thus:

$$\psi = \frac{2 \cdot A_{\text{flow}}}{A_{\text{enc}}} \cdot 100\% = \frac{A_{\text{flow}}}{A_{\text{enc}}} \cdot 200\%, \tag{5.31}$$

with the perforation coefficient ψ in percent, the surface area A_{flow} of the effective airflow, and the overall enclosure surface A_{enc}.

The parameter A_{flow} is the *effective* overall cross section of the flow channels. If only the bottom is perforated, no airstream can form, and $A_{\text{flow}} = 0$, whereas $A_{\text{flow}} = 25$ cm^2 for an air inlet area of 25 cm^2 and an exhaust area of 35 cm^2. The intake and outlet vents can also be located in the side panels (Fig. 5.27). Perforations and ventilation slots (side louvers) are equivalent for the same cross sections.

A perforation coefficient $\psi = 20\%$ means that 10% of each of the top and bottom surfaces (or side surfaces) are perforated [note the factor 2 in Eq. (5.31)]. The perforation coefficient should not exceed 25%, as there are no additional benefits to be gained with higher values [6].

We can estimate the mean overtemperature ΔT of the enclosure surface with respect to the surrounding, ambient air based on the perforation coefficient ψ with the help of graphs such as shown in Fig. 5.28. To illustrate, we investigated unpainted enclosures ($\varepsilon = 0.05$) with a very low heat dissipation rate through radiation to ensure a good margin of error with the overtemperature measurements in the graph.

Knowing the overtemperature ΔT of the enclosure surface with respect to the surrounding air, the thermal resistance between the perforated enclosure surface and ambient can be calculated using Eq. (5.30).

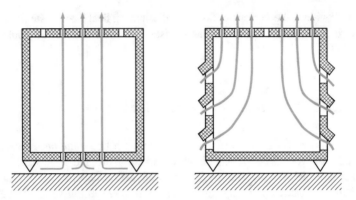

Fig. 5.27 Examples of open enclosures with intake and exhaust vents. *Note* the excessive (unused) exhaust vents that do not contribute to the airflow in both examples

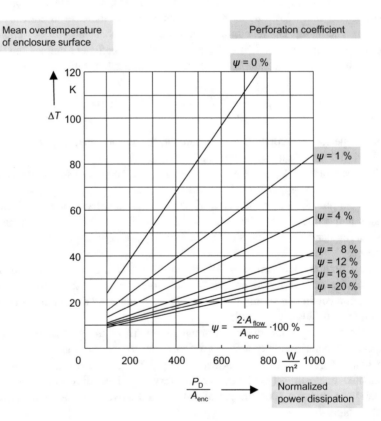

Fig. 5.28 Rise of enclosure temperature above the ambient air temperature (mean overtemperature ΔT) for unpainted enclosure surfaces ($\varepsilon = 0.05$) with different perforations as a function of the normalized input power (power dissipation) of the system [6]

Fig. 5.29 System enclosure
used in Examples 5.1 and 5.2

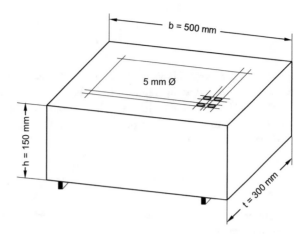

Example 5.1 The electronic components in an enclosure depicted in Fig. 5.29
dissipate 155 W of power. The dimensions of the enclosure are width $W = 0.5$ m =
500 mm, height $H = 0.15$ m = 150 mm, and depth $D = 0.3$ m = 300 mm. The
enclosure is unpainted ($\varepsilon = 0.05$) with a very low heat dissipation rate through
radiation, with thermal properties as illustrated in Fig. 5.28.

- What is the mean overtemperature ΔT of the enclosure with respect to the
  surrounding air if there are no perforations in the enclosure panels?
- How many round cut-outs with a diameter $d = 5$ mm are needed in the enclosure panels to ensure the mean overtemperature does not exceed $\Delta T = 30$ K?
- What are the thermal resistance values R_{th} of the sealed and open enclosures?

$$
\begin{aligned}
A_{enc} &= 2 \cdot (W \cdot H + W \cdot D + H \cdot D) \\
&= 2 \cdot (0.5 \cdot 0.15 + 0.5 \cdot 0.3 + 0.15 \cdot 0.3)\, \mathrm{m}^2 = 0.54\, \mathrm{m}^2, \\
\frac{P_D}{A_{enc}} &= \frac{155\,\mathrm{W}}{0.54\,\mathrm{m}^2} = 287\, \frac{\mathrm{W}}{\mathrm{m}^2}.
\end{aligned}
$$

We can read off a value of $\Delta T = 50$ K for the overtemperature of the sealed
enclosure ($\psi = 0\%$) in Fig. 5.28. Also shown is that a perforation coefficient
$\psi = 1\%$ is required to maintain a mean overtemperature $\Delta T = 30$ K. This yields the
area of the effective airflow

$$
A_{flow} = \frac{\psi}{200\%} \cdot A_{enc} = \frac{1\%}{200\%} \cdot 0.54\, \mathrm{m}^2 = 27 \cdot 10^{-4}\, \mathrm{m}^2.
$$

The number of cut-outs corresponding to the effective overall cross section of the
flow channels is:

$$z = \frac{A_{\text{flow}}}{d^2 \cdot \frac{\pi}{4}} = \frac{27 \cdot 10^{-4}\, \text{m}^2}{(0.005\, \text{m})^2 \cdot \frac{\pi}{4}} = 138.$$

Hence, the top and bottom of the enclosure need to have 138 holes *each* in order to maintain an overtemperature $\Delta T \leq 30$ K.

Finally, let us calculate how much perforation reduces the thermal resistance of the enclosure to ambient. The mean thermal resistance between the sealed enclosure surface and ambient is:

$$R_{\text{th}} = \frac{\Delta T}{P_{\text{D}}} = \frac{50\, \text{K}}{155\, \text{W}} = 0.32 \frac{\text{K}}{\text{W}}.$$

The thermal resistance between the open (perforated) enclosure surface and the surroundings is:

$$R_{\text{th\_p}} = \frac{\Delta T}{P_{\text{D}}} = \frac{30\, \text{K}}{155\, \text{W}} = 0.19 \frac{\text{K}}{\text{W}}.$$

Hence, with 1% of the enclosure effectively perforated, the thermal resistance between enclosure and ambient can be reduced by almost 50%.

5.5.5   Heat Dissipation from Sealed Enclosures

Recall that the overtemperature of the surface of a sealed enclosure with respect to the ambient air is not a function of the enclosure inside temperature, but is the product of the heat to be transferred and the thermal resistance between enclosure and surroundings [see Eq. (5.30)].

Heat is dissipated from the enclosure surface to the surroundings by means of convection and radiation. We can make a good engineering approximation that the system enclosure is fully enclosed by the room walls and that their surface temperature T_{wall} is approximately the same as the ambient air temperature T_{air}, which we assume to be constant. This common temperature is called the ambient temperature T_{a} below. The heat transfer coefficient for the convection h_{c} should be assigned as per Table 5.9, while the heat transfer coefficient h_{r} for the radiation can be determined by the chart in Fig. 5.15.

The heat P_{D} dissipated by convection and radiation can then be expressed as follows:

$$P_{\text{D}} = P_{\text{conv}} + P_{\text{rad}}, \tag{5.32a}$$

$$P_{\text{D}} = h_{\text{c}} \cdot A_{\text{conv}} \cdot (T_{\text{enc}} - T_{\text{a}}) + h_{\text{r}} \cdot A_{\text{rad}} \cdot (T_{\text{enc}} - T_{\text{a}}), \tag{5.32b}$$

where P_{conv} denotes the heat dissipated by convection in W, P_{rad} denotes the heat dissipated by radiation in W, h_c denotes the heat transfer coefficient for the convection in W/(m² K) (see Table 5.9), A_{conv} denotes the effective surface area for convection in m², T_{enc} denotes the mean temperature of the enclosure surface, T_a denotes the ambient temperature, h_r denotes the heat transfer coefficient for the radiation in W/(m² K) (see Fig. 5.15), and A_{rad} denotes the effective surface area for the radiation in m².

The enclosure overtemperature, i.e., the mean temperature differential of the enclosure to its surrounding air, is expressed as follows:

$$\Delta T = T_{enc} - T_a = \frac{P_D}{h_c \cdot A_{conv} + h_r \cdot A_{rad}} = P_D \cdot R_{th}. \tag{5.32c}$$

The resulting thermal resistance between the enclosure surface and the surroundings is:

$$R_{th} = \frac{1}{h_c \cdot A_{conv} + h_r \cdot A_{rad}}. \tag{5.32d}$$

Example 5.2 A power input $P_D = 400$ W is evenly dissipated to all sides in a system enclosure whose dimensions are as follows: width $W = 1$ m, height $H = 1$ m, and depth $D = 0.5$ m. The enclosure is installed so that heat can be dissipated by natural convection and radiation via the floor as well. The air temperature in the room and the temperature of the walls is 20 °C and thus $T_a = 20$ °C. The emissivity ε of the painted enclosure surface is 0.85.

- What is the mean surface temperature T_{enc} of the enclosure?
- What is the thermal resistance R_{th} between enclosure surface and the surroundings?

The enclosure is divided into vertical and horizontal surfaces with the associated heat transfer coefficients to calculate the heat dissipation by convection (see Table 5.9).
Vertical panel surfaces are:

$$A_{conv\_sides} = 2(W \cdot H + D \cdot H) = 2 \cdot (1 \cdot 1 + 0.5 \cdot 1)\, m^2 = 3\, m^2,$$

Top horizontal panel surface is:

$$A_{conv\_top} = W \cdot D = 1\, m \cdot 0.5\, m = 0.5\, m^2,$$

Bottom horizontal panel surface is:

$$A_{\text{conv\_bot}} = W \cdot D = 1 \text{ m} \cdot 0.5 \text{ m} = 0.5 \text{ m}^2.$$

The entire enclosure surface that radiates heat is:

$$A_{\text{rad}} = 2(W \cdot H + D \cdot H + W \cdot D) = 2(1 \cdot 1 + 0.5 \cdot 1 + 1 \cdot 0.5) \text{ m}^2 = 4 \text{ m}^2.$$

The heat dissipation can thus be expressed as:

$$\begin{aligned}
P_{\text{D}} = {} & h_{\text{c\_sides}} \cdot A_{\text{conv\_sides}} \cdot (T_{\text{enc}} - T_{\text{a}}) + h_{\text{c\_top}} \cdot A_{\text{conv\_top}} \cdot (T_{\text{enc}} - T_{\text{a}}) \\
& + h_{\text{c\_bot}} \cdot A_{\text{conv\_bot}} \cdot (T_{\text{enc}} - T_{\text{a}}) \\
& + h_{\text{r}} \cdot A_{\text{rad}} \cdot (T_{\text{enc}} - T_{\text{a}}).
\end{aligned}$$

The mean enclosure overtemperature with respect to the surroundings is:

$$\begin{aligned}
\Delta T = T_{\text{enc}} - T_{\text{a}} &= \frac{P_{\text{D}}}{h_{\text{c\_sides}} \cdot A_{\text{conv\_sides}} + h_{\text{c\_top}} \cdot A_{\text{conv\_top}} + h_{\text{c\_bot}} \cdot A_{\text{conv\_bot}} + h_{\text{r}} \cdot A_{\text{rad}}} \\
&= P_{\text{D}} \cdot R_{\text{th}}.
\end{aligned}$$

We cannot provide a sealed enclosure solution, as the heat transfer coefficient used in the calculation depends on the overtemperature we are looking for. We use an iteration procedure in this case to find the overtemperature: We use temperature estimates in our calculations and compare them with the calculated overtempera-tures until there is a sufficient agreement between estimated and calculated values.

First, we estimate the mean overtemperature as $\Delta T = 20$ K, i.e., $T_{\text{enc}} = 40$ °C. We can then calculate the following parameters:

$$h_{\text{c\_sides}} = 1.57 \cdot 20^{0.33} \frac{\text{W}}{\text{m}^2 \text{ K}} = 4.22 \frac{\text{W}}{\text{m}^2 \text{ K}},$$

$$h_{\text{c\_top}} = 1.3 \cdot 1.57 \cdot 20^{0.33} \frac{\text{W}}{\text{m}^2 \text{ K}} = 5.48 \frac{\text{W}}{\text{m}^2 \text{ K}},$$

$$h_{\text{c\_bot}} = 0.7 \cdot 1.57 \cdot 20^{0.33} \frac{\text{W}}{\text{m}^2 \text{ K}} = 2.95 \frac{\text{W}}{\text{m}^2 \text{ K}},$$

$$h_{\text{r}} = 0.85 \cdot 6.3 \frac{\text{W}}{\text{m}^2 \text{ K}} = 5.35 \frac{\text{W}}{\text{m}^2 \text{ K}},$$

from the equations for h_{c} in Table 5.9 (turbulent flow, as $\Delta T > (0.84/L)^3$; $c_{\text{tur}}(T_{\text{m}} = 30$ °C$) = 1.57$) and by determining h_{r} in Fig. 5.15.

The mean overtemperature is thus:

$$\Delta T = T_{enc} - T_a$$

$$= \frac{400\,\text{W}}{(4.22 \cdot 3 + 5.48 \cdot 0.5 + 2.95 \cdot 0.5 + 5.35 \cdot 4)\text{W}\,\text{m}^{-2} \cdot \text{K}^{-1}\,\text{m}^2}$$

$$= \frac{400\,\text{W}}{38.27\,\text{W}\,\text{K}^{-1}} = 10.45\,\text{K}.$$

We now perform a second calculation with the estimate $\Delta T = 10$ K ($T_{enc} = 30\,°\text{C}$) as there is quite a large difference between the estimated and the calculated temperature. Setting $c_{tur}(T_m = 25\,°\text{C}) = 1.59$ from Table 5.9, we obtain:

$$h_{c\_sides} = 1.59 \cdot 10^{0.33}\,\frac{\text{W}}{\text{m}^2\,\text{K}} = 3.40\,\frac{\text{W}}{\text{m}^2\,\text{K}},$$

$$h_{c\_top} = 1.3 \cdot 1.59 \cdot 10^{0.33}\,\frac{\text{W}}{\text{m}^2\,\text{K}} = 4.42\,\frac{\text{W}}{\text{m}^2\,\text{K}},$$

$$h_{c\_bot} = 0.7 \cdot 1.59 \cdot 10^{0.33}\,\frac{\text{W}}{\text{m}^2\,\text{K}} = 2.38\,\frac{\text{W}}{\text{m}^2\,\text{K}},$$

$$h_r = 0.85 \cdot 6.0\,\frac{\text{W}}{\text{m}^2\,\text{K}} = 5.1\,\frac{\text{W}}{\text{m}^2\,\text{K}}.$$

$$\Delta T = \frac{400\,\text{W}}{(3.40 \cdot 3 + 4.42 \cdot 0.5 + 2.38 \cdot 0.5 + 5.1 \cdot 4)\text{W}\,\text{m}^{-2} \cdot \text{K}^{-1}\,\text{m}^2}$$

$$= \frac{400\,\text{W}}{34.0\,\text{W}\,\text{K}^{-1}} = 11.76\,\text{K}$$

$$T_{enc} = T_a + \Delta T = 31.76\,°\text{C}.$$

We not do need to run another iteration, as the estimated overtemperature $\Delta T = 10$ K matches the calculated overtemperature ($\Delta T = 11.76$ K) quite well.

We should point out that this is a surface *mean* overtemperature. The assumption in this example that heat dissipation is uniform to all sides does not generally hold in practice.

The overtemperature calculation yields the mean thermal resistance between enclosure surface and surroundings as:

$$R_{th} = \frac{1}{h_{c_{sides}} \cdot A_{conv_{sides}} + h_{c_{top}} \cdot A_{conv_{top}} + h_{c_{bot}} \cdot A_{conv_{bot}} + h_r \cdot A_{rad}}$$

$$= \frac{1}{34.0\,\text{W}\,\text{K}^{-1}} = 0.029\,\frac{\text{K}}{\text{W}}$$

as per Eq. (5.32d).

Some 60%, i.e., the majority of the power input $P_D = 400$ W, is dissipated by radiation:

$$P_{rad} = h_r \cdot A_{rad} \cdot \Delta T = 5.1\,\text{W}\,\text{m}^{-2}\,\text{K}^{-1} \cdot 4\text{m}^2 \cdot 11.76\,\text{K} = 240\,\text{W}.$$

Example 5.3 Consider a system enclosure with dimensions width $W = 0.5$ m $= 500$ mm, height $H = 0.15$ m $= 150$ mm, and depth $D = 0.3$ m $= 300$ mm (see Fig. 5.29). Calculate the power P_D that can be dissipated from the enclosure surface:

(a) unpainted (bare) and sealed surface ($\varepsilon_a = 0.05$);
(b) painted (finished) and sealed surface ($\varepsilon_b = 0.85$);
(c) unpainted, perforated surface ($\varepsilon_c = 0.05$), if 12% of the entire surface area (6% surface for air intake and 6% for air exhaust) is perforated; and
(d) perforated surface, (as in c), but enclosure is painted ($\varepsilon_d = 0.85$).

The ambient temperature is $T_a = 20$ °C, and the approved overtemperature of the enclosure surface is $\Delta T = 15$ K. Assume uniform heat dissipation upward and through the side panels (there is no heat dissipation through the base plate). Therefore, the effective surface area for radiation is:

$$\begin{aligned}
A_{rad} &= W \cdot D + 2(H \cdot D + H \cdot W) \\
&= 0.5\,\text{m} \cdot 0.3\,\text{m} + 2(0.15 \cdot 0.3 + 0.15 \cdot 0.5)\text{m}^2 = 0.39\,\text{m}^2.
\end{aligned}$$

The effective surface area for convection is:

$$\begin{aligned}
A_{conv\_sides} &= 2(W \cdot H + D \cdot H) = 2(0.5 \cdot 0.15 + 0.3 \cdot 0.15)\text{m}^2 = 0.24\,\text{m}^2 \\
A_{conv\_top} &= W \cdot D = 0.5\,\text{m} \cdot 0.3\,\text{m} = 0.15\,\text{m}^2.
\end{aligned}$$

(a) We obtain h_r from Fig. 5.15 and h_c from Table 5.9 (laminar flow, as $\Delta T \leq (0.84/L)^3$; $c_{lam}(T_m = 27.5$ °C$) \approx 1.37$) for the unpainted, sealed enclosure as follows:

$$h_r = \varepsilon_1 \cdot h_r^* = 0.05 \cdot 6.2\,\frac{\text{W}}{\text{m}^2\,\text{K}} = 0.31\,\frac{\text{W}}{\text{m}^2\,\text{K}}$$

$$h_{c\_sides} = 1.37 \cdot \sqrt[4]{\frac{\Delta T/\text{K}}{L/\text{m}}} = 1.37 \cdot \sqrt[4]{\frac{15}{0.15}} = 4.3\,\frac{\text{W}}{\text{m}^2\,\text{K}}$$

$$h_{c\_top} = 1.3 \cdot 1.37 \cdot \sqrt[4]{\frac{\Delta T/\text{K}}{L_{min}/\text{m}}} = 1.3 \cdot 1.37 \cdot \sqrt[4]{\frac{15}{0.3}} = 4.7\,\frac{\text{W}}{\text{m}^2\,\text{K}}$$

$$\begin{aligned}
P_D &= (h_r \cdot A_{rad} + h_{c\_sides} \cdot A_{conv\_sides} + h_{c\_top} \cdot A_{conv\_top}) \cdot \Delta T \\
&= (0.31 \cdot 0.39 + 4.3 \cdot 0.24 + 4.7 \cdot 0.15)\,\frac{\text{W}}{\text{m}^2\,\text{K}} \cdot \text{m}^2 \cdot 15\,\text{K} = 27.87\,\text{W}\,.
\end{aligned}$$

This means that 26.06 W (93%) total heat output is dissipated by convection and 1.81 W (7%) by radiation. Note that we can increase the level of radiation to approximately 30% by using a slightly oxidized surface (with a higher emissivity $\varepsilon = 0.25$) instead of the bare surface ($\varepsilon = 0.05$).

(b) The radiation heat transfer coefficient $h_r = 5.27$ W/(m^2 K) for a sealed, painted enclosure leads to a total approved dissipated heat output $P_D = 56.88$ W. The heat dissipation to the surroundings is 26.06 W (46%) convection and 30.82 W (54%) radiation. The use of a finished surface in this example illustrates an important point: The approved power dissipation can be almost doubled *at constant overtemperature* by finishing (painting) the enclosure and thus increasing the radiation exchange!

(c) According to Eq. (5.31), the perforation coefficient for the perforated enclosure is:

$$\psi = \frac{2 \cdot A_{\text{flow}}}{A_{\text{enc}}} \cdot 100\% = \frac{A_{\text{flow}}}{A_{\text{enc}}} \cdot 200\% = 12\%,$$

i.e., 6% of each of the top and bottom surfaces are perforated.
For $\Delta T = 15$ K and $\psi = 12\%$, we can read off the normalized power dissipation of our system from the graph in Fig. 5.28 as

$$\frac{P_D}{A_{\text{enc}}} = 280 \frac{\text{W}}{\text{m}^2}.$$

Applying the entire enclosure surface area $A_{\text{enc}} = 0.54$ m^2, we calculate the approved power dissipation:

$$P_D = 0.54 \text{ m}^2 \cdot 280 \frac{\text{W}}{\text{m}^2} = 151 \text{ W}.$$

The addition of perforations in this example illustrates another important point: We can increase the dissipated heat (power) by a factor of approximately 5.4 (151 W vs. 27.87 W) with a 12% perforation of the enclosure!

(d) Finishing, e.g., painting the perforated enclosure increases the emissivity from $\varepsilon_c = 0.05$ to $\varepsilon_d = 0.85$ (h_r^* again from Fig. 5.15). Recall that the enclosure does not dissipate any heat through the base plate. If we ignore the radiation from the bare (unpainted) enclosure and the reduction in surface area by perforation (heat is radiated through cut-outs as well), the additional radiation obtained with the finished surface is:

$$P_{\text{rad}} = A_{\text{rad}} \cdot (\varepsilon_d \cdot h_r^*) \cdot \Delta T = 0.39 \text{ m}^2 \cdot (0.85 \cdot 6.2 \text{ W m}^{-2} \text{ K}^{-1}) \cdot 15 \text{ K}$$
$$= 30.8 \text{ W}.$$

The total heat (power) dissipated from a perforated *and* finished enclosure is:

$$P_D = 151\ \text{W} + 30.8\ \text{W} = 181.8\ \text{W}.$$

By combining thermal management techniques in this example, we illustrate an important further concept: We have effectively increased total heat output compared to a sealed and bare (unpainted) enclosure (a) by a factor of 6.5 (181.8 W vs. 27.87 W) using perforation (c) and surface finishing (d).

5.5.6 Heat Transfer Through Enclosure Panels

Heat from electronic components is absorbed by enclosure panels by convection and radiation, thermally conducted through the enclosure panels, and then released by convection and radiation. If heat sources are connected directly, or via substrates, supports, etc., with the panel, heat will also be transported to the panels via conduction. Note that heat can only be dissipated from electronic components to panels by radiation if the panels are in direct radiation exchange with them.

Cabinet and enclosure panels are typically coated with 0.03–0.05-mm-thick paint finishes. We have already dealt with the improvements in emissivity of bare metallic surfaces ($\varepsilon = 0.02$–0.25) by finishing ($\varepsilon = 0.8$–0.95). We will now examine the impact of finishes on heat transfer through the enclosure panel.

Example 5.4 Consider the configuration in Example 5.2, where heat of $P_D = 400$ W is evenly dissipated through all sides of an enclosure. We calculated the total surface area of the enclosure at $A_{\text{enc}} = 4\ \text{m}^2$ and the mean surface temperature $T_{\text{enc\_1}}$ of the enclosure at 31.76 °C (overtemperature 11.76 K). The enclosure consists of $d_1 = 1$-mm-thick steel sheet ($k_1 = 59$ W/(m K), see Table 5.6), and a $d_2 = 0.05$-mm finish on both sides [$k_2 = 0.2$ W/(m K)]. The thermal parameters and flows are depicted in the schematic in Fig. 5.30.

(a) What is the temperature differential between the internal and external panel surfaces?
(b) What is the proportion of the finishes in the temperature differential?

(a) The thermal resistance of the steel sheet is:

$$R_{\text{cond\_1}} = \frac{d_1}{k_1 \cdot A_{\text{enc}}} = \frac{0.001\,\text{m}}{59\,\frac{\text{W}}{\text{m·K}} \cdot 4\,\text{m}^2} = 4.24 \cdot 10^{-6}\,\frac{\text{K}}{\text{W}}.$$

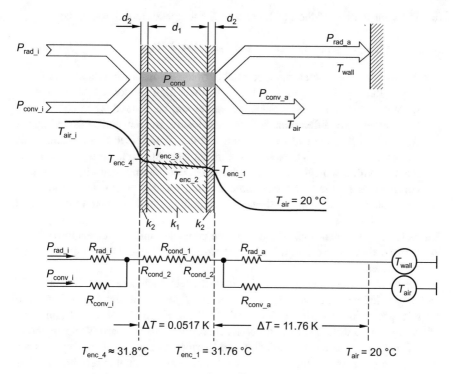

Fig. 5.30 Heat transfer through enclosure panel finished on both sides (Example 5.4)

The thermal resistance of a paint finish is:

$$R_{\text{cond\_2}} = \frac{d_2}{k_2 \cdot A_{\text{enc}}} = \frac{50 \cdot 10^{-6}\,\text{m}}{0.2\frac{\text{W}}{\text{m·K}} \cdot 4\,\text{m}^2} = 62.5 \cdot 10^{-6}\frac{\text{K}}{\text{W}}.$$

The total thermal resistance of the panel finished on both sides is:

$$R_{\text{cond}} = R_{\text{cond\_1}} + 2 \cdot R_{\text{cond\_2}} = 129.24 \cdot 10^{-6}\frac{\text{K}}{\text{W}},$$

where the two finishing coats contribute to 97% of the total resistance. Therefore, the temperature differential between internal and external panel surface is:

$$T_{\text{enc\_4}} - T_{\text{enc\_1}} = R_{\text{cond}} \cdot P_{\text{D}} = 129.24 \cdot 10^{-6}\frac{\text{K}}{\text{W}} \cdot 400\,\text{W} = 0.0517\,\text{K}\,.$$

Hence, the temperature of the inner panel surface is approximately 31.8 °C and, thus, does not significantly differ from the outer surface temperature of 31.76 °C (calculated in Example 5.2).

(b) The contribution of the finish coats to the temperature differential is:

$$(T_{enc\_4} - T_{enc\_3}) + (T_{enc\_2} - T_{enc\_1}) = 2 \cdot R_{cond\_2} \cdot P_D = 0.05 \text{ K}.$$

Almost all the temperature differential between the inside and outside surface is across the layers of paint finish due to their high proportion of the thermal resistance. Nevertheless, these small values can be ignored when compared to the temperature differential between enclosure surface and surroundings of $\Delta T = 11.76$ K generated by radiation and convection (see Example 5.2) (this means $T_{enc\_4} \approx T_{enc\_1}$). Note that this temperature differential can only be ignored with low heat fluxes. Finally, when considering heat dissipation from components *by thermal conduction*, thermal resistances and temperature differentials across surface finishes, such as paints, should be taken into account.

Example 5.5 Our final application example considers multi-layer enclosures with air gaps as they are common for shielding against high-energy, high-frequency electromagnetic fields.

The inside surface of the $d_1 = 1$ mm steel sheet enclosure from Examples 5.2 and 5.4 is clad with a $d_4 = 1$ mm copper sheet ($k_4 = 372$ W/(m·K), see Table 5.6) for ESD protection. An *air gap* ($k_3 = 0.0257$ W/(m·K) 20 °C) of $d_3 = 2$ mm is part of the design as well. There is no convection owing to the narrowness of the air gap. The heat exchange by radiation and thermal conduction across spacers and fixing elements between the two panels is ignored as an initial approximation. The design is shown in the schematic in Fig. 5.31.

What is the temperature differential $T_{enc\_6} - T_{enc\_1}$ between the external surfaces of the two panels and what proportion of the differential is due to the air gap?

According to Example 5.4, the temperature differential between inside and outside surfaces of the steel panel is $T_{enc\_4} - T_{enc\_1} = 0.0517$ K. The total surface areas of the steel and copper panel are approximately equal. First, let us determine the thermal resistance of the air gap with:

$$R_{cond\_3} = \frac{d_3}{k_3 \cdot A_{enc}} = \frac{0.002 \text{ m}}{0.0257 \frac{\text{W}}{\text{m·K}} \cdot 4 \text{ m}^2} = 19.5 \cdot 10^{-3} \frac{\text{K}}{\text{W}}.$$

This yields a temperature differential across the air gap of:

$$T_{enc\_5} - T_{enc\_4} = R_{cond\_3} \cdot P_D = 0.0195 \frac{\text{K}}{\text{W}} \cdot 400 \text{ W} = 7.8 \text{ K}.$$

Next, the values for the thermal resistance and the temperature differential of the copper plate are:

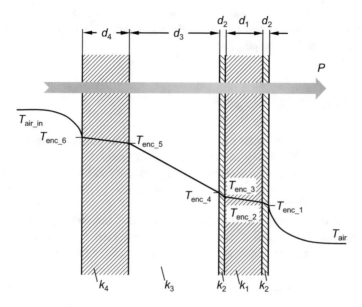

Fig. 5.31 Heat transfer through a double panel with air layer in Example 5.5

$$R_{\text{cond\_4}} = \frac{d_4}{k_4 \cdot A_{\text{enc}}} = \frac{0.001 \,\text{m}}{372 \dfrac{\text{W}}{\text{m} \cdot \text{K}} \cdot 4 \,\text{m}^2} = 0.67 \cdot 10^{-6} \frac{\text{K}}{\text{W}},$$

$$T_{\text{enc\_6}} - T_{\text{enc\_5}} = R_{\text{cond\_4}} \cdot P_{\text{D}} = 0.67 \cdot 10^{-6} \frac{\text{K}}{\text{W}} \cdot 400 \,\text{W} = 0.00027 \,\text{K}.$$

The total temperature differential across the double panel is thus:

$$T_{\text{enc\_6}} - T_{\text{enc\_1}} = 0.0517 \,\text{K} + 7.8 \,\text{K} + 0.00027 \,\text{K} \approx 7.85 \,\text{K}.$$

Obviously, almost all of the total temperature differential, 7.8 K, is across the air gap.

If we add this temperature differential ΔT_{enc} = 7.85 K across the enclosure to the overtemperature $\Delta T_{\text{enc\_air}}$ = 11.76 K of the outside surface from Example 5.2, we obtain an internal surface temperature that is 19.61 K above ambient air. Hence, the mean internal surface temperature of the copper plate is:

$$T_{\text{enc\_6}} = 19.61 \,\text{K} + 20 \,^\circ\text{C} = 39.61 \,^\circ\text{C}.$$

Due to the air gap, this temperature of the inner enclosure surface is significantly hotter than the inner surface of the steel panel calculated in Example 5.4 (31.8 °C).

The examples show that the major temperature differentials for heat transfer through enclosures occur as a result of convection and radiation at the inside and external surfaces. The thermal conduction resistances for the enclosure panels and the finishing coats can be ignored. Enclosures that contain thin layers of air, in which the heat transfer occurs only by thermal conduction (static air), should be absolutely avoided where possible.

5.5.7 Interior Air Temperatures

The air temperature inside the enclosure is related to the surface temperature of the inside of the enclosure panel, as are all the other temperatures inside the system, all the way to the internal temperature (e.g., junction temperature T_j) of the heat generating electronic component. This relationship can be illustrated by calculating the interior air temperature next.

If we assume the mean heat transfer coefficient at the inside panel due to free convection is $\alpha_K = 5$ W/(m^2 K), we can express the mean overtemperature $\Delta T_{enc\_air\_in}$ between the inside surface and the air in the enclosure as

$$\Delta T_{enc\_air\_in} = T_{enc} - T_{air\_in} = R_{conv} \cdot P_D,$$

$$R_{conv} = \frac{1}{h_c \cdot A_{enc}} = \frac{1}{5 \text{ W m}^{-2}\text{K}^{-1} \cdot 4 \text{ m}^2} = 0.05 \frac{K}{W},$$

$$\Delta T_{enc\_air\_in} = 0.05 \frac{K}{W} \cdot 200 \text{ W} = 10 \text{ K}.$$

The power of 200 W is based on the assumption that half the power dissipation ($P_D = 400$ W) of the electronic component is transmitted by convection and the other half by radiation to the inside of the enclosure.[3] The radiation portion can be ignored in this calculation since we assume that the component's radiation does not affect the internal air temperature.

Using the calculated overtemperature between internal air and enclosure inside panel of 10 K, and taking into account the temperature differential across the enclosure panel, which was previously determined, the mean interior air temperature of the electronic system is:

$$T_{air\_in} = T_{enc\_4} + \Delta T_{enc\_air\_in} = 31.8\,^{\circ}\text{C} + 10\,\text{K} = 41.8\,^{\circ}\text{C},$$

for Example 5.4 (steel enclosure),
    and

$$T_{air\_in} = T_{enc\_6} + \Delta T_{enc\_air\_in} = 39.61\,^{\circ}\text{C} + 10\,\text{K} = 49.61\,^{\circ}\text{C},$$

for Example 5.5 (steel and copper enclosure with air gap).

---

[3]While bare surfaces radiate approximately 25% of the dissipated heat (i.e., 75% by convection), painted surfaces radiate roughly 55–60% (i.e., 40–45% by convection).

Based on these interior air temperatures, the overtemperature(s) of the components (all the way to the junction temperature T_j) can be calculated using the same method (see also Sect. 5.5.1).

5.5.8 Heat Transfer Inside an Open Enclosure

It is obvious that the internal temperature of electronic systems (see previous Sect. 5.5.7) is higher than the panel temperatures. The heat transfer processes internal to an enclosure panel are much more complex than for heat transfer through the enclosure panel (see Sect. 5.5.6) or for the heat transfer between the external enclosure surface and surrounding ambient (see Sects. 5.5.4 and 5.5.5).

There is often insufficient space between panels, modules, and components, to allow the airflow necessary for convection. In such cases, heat can only be transferred to the enclosure panel by radiation from heat sources that are in direct radiation exchange with the panel. Heat transfer or dissipation by thermal conduction is possible if the heat sources are connected in a thermally conductive manner with the enclosure panel. There are often design issues which do not allow this, however. Dissipating heat by convection is very cost-effective if the required measures are put in place at the design stage.

Engineers usually resort to empirical techniques, such as measurements on models and prototypes, to calculate the interior temperatures, as analytical methods are very inaccurate due to the complex configurations encountered in practice. The use of rough calculations of the mean internal temperature in a system, such as carried out in Sect. 5.5.7, is approved, if convection flows can easily form at the panels and the heat can be removed evenly to all sides.

In the following, we will explore some of the problems encountered with heat transfer and dissipation inside an electronic system with an open enclosure, such as shown in the schematic in Fig. 5.32. We assume that the system contains plug-in subracks, in which a number of printed circuit boards are inserted vertically. The heat to be removed is produced by the printed circuit boards. The heat is mostly transported by convection. We recommend the use of natural convection if possible for reliability reasons. The thermal conditions are mainly determined by the *distance between the printed circuit boards*. The distance between vertical, heated boards should be two times the velocity boundary layer thickness, which is approximately 1–2 cm for boards up to 50 cm in height, according to Sect. 5.3.3. Tests have shown that increasing the board clearance to over 2 cm brings no significant reduction in overtemperatures. This applies to printed circuit boards with components placed on one side only (single-sided PCBs). A clearance >3 cm is required for double-sided printed circuit boards; 1–2 cm is adequate for forced convection.

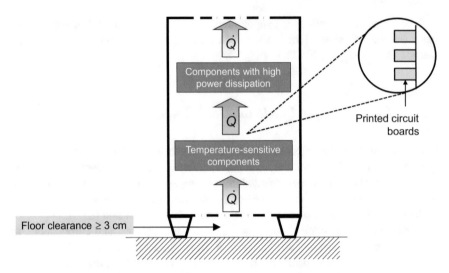

Fig. 5.32 Convection in an open electronic system with subracks, containing, for example, vertically mounted printed circuit boards

It is recommended that plug-in modules with temperature-sensitive components be inserted at the bottom of the enclosure and that modules with components emitting most heat be inserted nearest the top of the enclosure. The effective overall cross section of the airstream should be maximized. Try to avoid intake and exhaust vents of different sizes. The legs supporting the enclosure should be at least 3 cm high to ensure that the air can easily flow into the system through the air inlets in the bottom panel.

Warm air can be blocked from rising into the upper modules by inserting inclined *baffles* if the overtemperature in the upper part of the system is too high due to excessive heat dissipation (Fig. 5.33 left). A sufficient number of cut-outs should be made in the front panels and the back panel to facilitate these baffles.

Some design solutions use the *chimney effect* (or stack effect) to dissipate heat from power supplies (Fig. 5.33 right). This effect is a special case of natural convection, where rising warm air is fully enclosed by vertical panels. The chimney "draft" is caused by the density difference between the heated air column in the chimney and an air column of the same height at outside temperature. The guide panels offer much less resistance to the rising warm air than the cold ambient air does, resulting in a brisk upwind that increases cooling significantly.

The circle and the square are the most suitable cross sections for flow. Rectangular cross sections are generally the only ones available in practice. Here, the smallest rectangle side should not be smaller than 3 cm to prevent the wall friction from obstructing the chimney effect too greatly. Care should be taken that the rising warm air is not obstructed with cables or fixed elements.

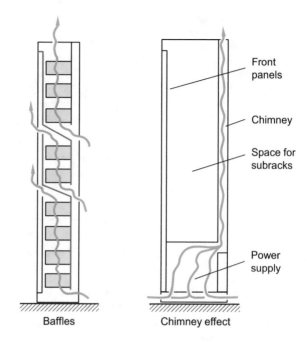

Fig. 5.33 Guiding the temperature distribution with baffles (*left*) and using the chimney effect to increase the cool airflow across the back panel (on the *right*)

Baffles   Chimney effect

5.5.9 Heat Transfer Inside a Sealed Enclosure

Sealed enclosures offer many advantages regarding issues such as shock protection and protection against ingress of foreign bodies or water and electromagnetic compatibility (EMC). Often, only sealed enclosures can be used in outdoor installations, as these systems are exposed to humidity, corrosive gases, dust, or dirt. If, however, natural convection is used for heat transfer and heat dissipation inside these enclosures, only low heat flux densities should occur (see Table 5.11).

Flow conditions in sealed enclosures are complex due to the different temperature distributions: Circulations arise due to rising and falling neighboring flows. Eddy zones that are almost imposable to describe mathematically can, due to the temperature gradients, arise in vertical spaces, for example, between enclosure panel and modules. Similar circulations occur in horizontal spaces, for example, in the upper part of the enclosure, if the bottom of the space is warmer than the top part. Therefore, single-sided printed circuit boards should never be fitted horizontally with the components facing downwards.

Conditions are even more challenging at the enclosure bottom. If the warmer surface is on top in a horizontal space, convection can hardly take place. There is only a small moving air stratification due to the density differences between the warmer board top and cooler board bottom, and the heat can practically only be conducted downward through the air layers. This issue should be taken into account with horizontally arranged printed circuit boards.

Fig. 5.34 Combining the concept of a heat exchanger with forced convection by two fans significantly improves heat transfer out of the sealed enclosure

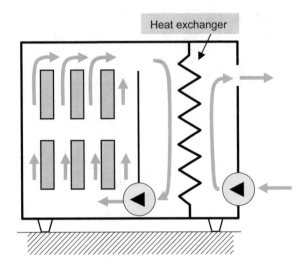

Two- to fourfold improvements in heat transfer rates can be achieved by installing a fan inside the enclosure to produce forced convection (see Table 5.11). Mechanical design measures are needed to produce good flow conditions in the enclosure to prevent "hot spots" from forming. Another significant improvement can be achieved by transferring heat produced inside an enclosure by thermal conduction to the enclosure panels.

Higher heat flux densities are allowed if forced convection is deployed both inside and outside sealed enclosures. The schematic in Fig. 5.34 shows how this can be achieved with a *heat exchanger* fitted in the system enclosure. The fan in the inner circuit circulates the heated inside air and thus provides airflow within the system. A second fan draws cool air in from the surroundings. The two airstreams flow past a large-surface-area heat exchanger that enables heat from the internal flow to be effectively transferred to the external flow, for removal from the system.

5.5.10 Forced Convection with Fans and Fan Selection

Recall that the heat transfer coefficients for gases are $h_c = (5–10)$ W/(m² K) for natural convection and $h_c = (10–120)$ W/(m² K) for forced convection (see Sect. 5.3.3, especially Fig. 5.9). The difference is caused by the enhanced velocity and thus higher flow rate of the fluid. Hence, by using a fan, we can remove heat at much higher rates and reduce the surface temperature of the components considerably better than with natural convection.

If the heat transfer or the "cooling problem" is caused by flat, heated boards, then the required mean flow velocity v can be roughly estimated with Eqs. (5.33a) and (5.33b). Equation (5.33a) expresses the heat transfer coefficient h_c as a function of the flow velocity v for forced laminar flow ($v < 5$ m/s) along a flat panel of length L, in any position and orientation, and Eq. (5.33b) expresses the conditions with forced turbulent flow ($v > 5$ m/s):

$$h_c \approx 3.9 \cdot \sqrt[2]{\frac{v / \frac{m}{s}}{L / m}}, \qquad\qquad (5.33a)$$

$$h_c \approx 5.9 \cdot \sqrt[4]{\frac{\left(v / \frac{m}{s}\right)^3}{L / m}}. \qquad\qquad (5.33b)$$

Generally speaking, the goal is to remove the power P_D from a given source of heat. Thus, we want to minimize the convection thermal resistance introduced in Sect. 5.3.3 with Eq. (5.12):

$$R_{conv} = \frac{1}{h_c \cdot A},$$

so that the overtemperature:

$$\Delta T = R_{conv} \cdot P_D$$

inside the heat source (e.g., a component) is as low as possible. If we want to dissipate heat from the component via a heat sink, as shown in Sect. 5.5.1, we need to reduce the thermal resistance between heat sink and surroundings in order to increase this heat transfer. Forced flow with fans is very effective here. The correlation between the reduction in the thermal resistance and the mean flow velocity v of the air is expressed with Eq. (5.34) and illustrated in Fig. 5.35. The trajectory of the curve $f_R = f(v)$ corresponds qualitatively to the correlation between h_c and v in Eq. (5.33a).

The thermal resistance R_{f_conv} in K/W with forced convection is expressed as follows:

$$R_{f\_conv} = f_R \cdot R_{conv}, \qquad\qquad (5.34)$$

Fig. 5.35 Correction factor f_R in Eq. (5.34) reduces the convection thermal resistance with increasing airflow velocity v provided by forced convection

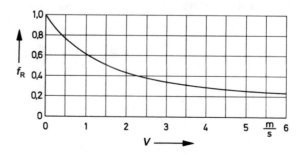

where f_R denotes the correction factor as per Fig. 5.35 and R_{conv} denotes the thermal resistance with natural convection as per Eq. (5.12) in K/W.

The airflow velocity in Fig. 5.35 and the cross-sectional area of the flow channel necessitate a volumetric flow rate $\dot{V}$ which can be calculated by:

$$\dot{V} = A \cdot v, \tag{5.35}$$

where $\dot{V}$ denotes the volumetric flow rate in m³/s, A denotes the cross section of the flow channel in m², and v denotes the mean flow velocity of the gas in m/s (see Fig. 5.35).

If heat is to be dissipated by a fan, the correlation between power dissipation, volumetric flow rate, and temperature differential between air intake and air exhaust can be expressed as follows:

$$P_D = \dot{V} \cdot \varrho \cdot c_p \cdot \Delta T_{air}, \tag{5.36}$$

where P_D denotes the heat dissipated by forced convection in W and $\dot{V}$ denotes the volumetric flow rate in m³/s from Eq. (5.35). The symbol ϱ denotes the air density in kg/m³; the following applies for dry air at standard pressure (sea level):

$$
\begin{array}{rll}
T = & 0\,°C: & \varrho = & 1.29\,kg/m^3 \\
& 20\,°C & & 1.2\,kg/m^3 \\
& 40\,°C & & 1.13\,kg/m^3 \\
& 60\,°C & & 1.06\,kg/m^3 \\
& 80\,°C & & 1.00\,kg/m^3 \\
& 100\,°C & & 0.95\,kg/m^3.
\end{array}
$$

Furthermore, c_p is the specific heat of air in W·s/(kg·K) [or J/(kg·K)] as a function of the air pressure or the elevation. The specific heat c_p for air at standard pressure, i.e., sea level, is:

$$c_p \approx 10^3 \cdot \frac{W \cdot s}{kg \cdot K}.$$

The parameter ΔT_{air} is the temperature differential of the air between inlet and outlet of the flow channel in K (i.e., how much hotter is the outgoing air compared to the incoming):

$$\Delta T_{air} = T_{outlet} - T_{inlet}. \tag{5.37a}$$

Using both temperatures, we can also calculate the mean inside air temperature T_{air_in} as follows:

$$T_{air\_in} \approx \frac{T_{outlet} + T_{inlet}}{2}. \tag{5.37b}$$

Equation (5.36) is the universal equation for heat transfer via mass flow; it yields the airflow required to dissipate a given amount of heat. Hence, it is widely used to calculate the heat dissipated by a fan.

There are four steps involved in selecting a fan for heat dissipation in an electronic system:

1. Calculate the required volumetric flow rate $\dot{V}$.
2. Establish the characteristic curve of the pressure drop for the system (system pressure curve).
3. Determine the operating points of candidate fans on the system pressure curve using their fan curves.
4. Select a fan based on the operating point that will deliver the necessary volumetric flow for the system.

Let us now discuss these four steps in more detail. First, we *calculate the required volumetric flow rate* $\dot{V}$ based on the known power input of our system P_D and an allowed temperature rise between air exhaust and air intake ΔT_{air}. This can be done either by rearranging Eq. (5.36) to express $\dot{V}$ or with the approximation formulas:

$$\dot{V} \, / \left(\frac{m^3}{s} \right) \approx \frac{1}{1200} \cdot \frac{P_D \, / \, W}{\Delta T_{air} \, / \, K}, \tag{5.38a}$$

or

$$\dot{V} \, / \left(\frac{m^3}{h} \right) \approx 3 \cdot \frac{P_D \, / \, W}{\Delta T_{air} \, / \, K}, \tag{5.38b}$$

with the heat to be removed P_D and the temperature differential ΔT_{air} of the air between the inlet and outlet of the flow channel.

Second, we must determine the pressure curve for the system to be cooled. Recall that every electronic system resists the air throughput generated by the fan, which produces a static pressure (also known as head loss) as a function of the generated volumetric flow. This system property is depicted by the *system pressure curve* (also: system impedance curve) introduced in Sect. 5.4.3 (Fig. 5.36).

The air resistance in the flow channel is composed of friction losses and baffle resistances. On the one hand, the curve can be generated by measuring the static pressure drop at different flow rates. On the other hand, we can calculate these losses for simple shapes only, which usually excludes electronic systems with their

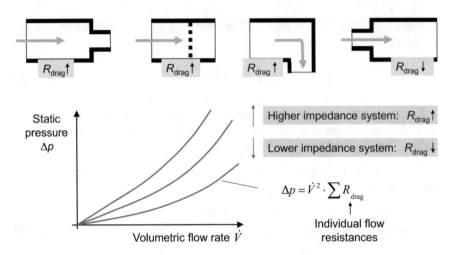

Fig. 5.36 Different system pressure curves depending on the air resistance R_{drag} in the flow channel. The "parabolic shape" of a system curve is due to the dependence of the pressure on the square of the volumetric flow rate; the static pressure increases quadratically with the airflow speed in fully turbulent flows

complex interior layout. Nevertheless, we wish to define the mathematical expression of static pressure Δp_{pipe} (head loss) in a pipe here, in order to be able to estimate the independent variables:

$$\Delta p_{pipe} = \frac{v^2 \cdot \varrho}{2}\left(\mu \cdot \frac{L}{D} + \Sigma \varsigma\right),$$ (5.39)

where Δp_{pipe} denotes the static pressure (friction loss) in a pipe in N/m^2, v denotes the mean flow velocity of air in m/s, ϱ denotes the density of air in kg/m^3, μ denotes the coefficient of friction between airflow and panel, L denotes the length of pipe, D denotes the inside pipe diameter, and ς the resistance coefficient for bends, cross-sectional changes and branches.

The friction loss (resistance) increases where the flow experiences directional and cross-sectional changes (see Fig. 5.36 top). The greater the change in flow, the higher the resistance coefficient ς. If, for example, we compare a sharp corner with a smooth bend, whose mean radius is six times the pipe cross section, the ratio of the resistance coefficients is on the order of 1.5:0.01 = 150:1. We should, therefore, be careful when using baffles to avoid sharp changes in direction.

There are less frictional losses in larger flow cross sections. The cross section should, however, be as small as possible at heat sources, as the heat transfer improves with increasing flow velocity.

Third, we need to relate our system pressure curve with the fan curve(s) for candidate fans. The *fan curve* introduced in Sect. 5.4.3 is normally determined experimentally by the manufacturer and supplied by them to help engineers select the right fan. It shows the specific relationship between the generated static pressure

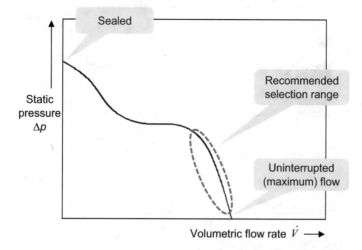

Fig. 5.37 Characteristic curve of a fan (fan curve) specifying the relation between its generated pressure and the volumetric flow rate. The latter can vary between no flow (system is sealed) and maximum flow (wide open system without resistance). The fan should be operated in the lower third of its pressure range to reduce noise

Δp and the volumetric flow rate $\dot{V}$ of the air for a fan (Fig. 5.37). The static pressure is generated by the fan to overcome the friction drag of air in the system and to produce the kinetic energy of the circulating air. The fan creates the highest pressure at zero flow rate (sealed system); the pressure decreases with the opening of the system.

Finally, we select the most appropriate fan based on its *operating point* which is the point of intersection of the fan curve with the system pressure curve (see Fig. 5.21). Specifically, the volumetric flow rate generated by a candidate fan for a given system is derived from the operating point. As illustrated in Fig. 5.38, a fan is selected based on the operating point that delivers sufficient volumetric flow for the system. As mentioned earlier, the operating point should be in the lower third of the pressure range to reduce noise.

If two similar fans are fitted in series, the pressures are added to provide an increased total pressure (Fig. 5.39 left). This method can be used to produce the required pressure with a number of fans so that the fans can run with less noise, for example. Vertex and inflection points on the curves remain unchanged. The volumetric flow rate does not change either.

If two similar fans are run in parallel, the volumetric flow rates are added for the same static pressure (Fig. 5.39 on the right). The characteristic curves should not have any vertex or inflection points so as to avoid undesirable effects, such as fluctuating volumetric flow rates or recirculation by one of the fans.

Air filters should be used to minimize dust contaminants piling up on the components. Fans can be fitted in the air intake, at the air exhaust or fitted in a bulkhead partition within the electronic system. A fan placed at the intake

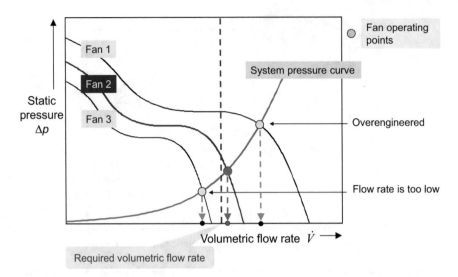

Fig. 5.38 To select a fan, various fan curves (here: fans 1, 2, and 3) are combined with the system pressure curve to obtain the fan operating points. These operating points define the volumetric flow rates generated by the fan for the respective system, which should correspond to the required volumetric flow rate calculated previously. In this example, fan 2 is chosen

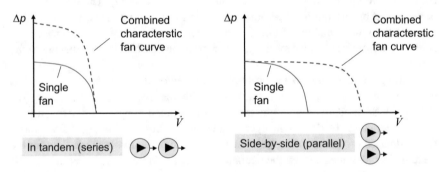

Fig. 5.39 Running two similar fans in series and in parallel increases the generated pressure (*left*) or the generated volumetric flow rate (on the *right*)

pressurizes the system and prevents air infiltration into the enclosure from other openings. The fan is also subjected to cool air, thereby increasing its reliability. However, the heat of the fan also adds up to the heat load of the system. "Suction" fans at the system air exhaust often lead to increased dust deposits inside the system, as the air not only enters exclusively through the inlet filter, but also through the openings in the enclosure.

Although the volumetric flow rate $\dot{V}$ is indifferent to the altitude of the installed fan, the lower density of air at higher altitudes leads to less air mass being delivered. (Air pressure is only 60% at 4000 m altitude compared to sea level, for example.)

Thus, the temperature rise of the cooling air will increase significantly, requiring a larger flow rate for compensation. This necessitates using Eq. (5.36) with well-established values for air density ϱ and specific heat c_p. Thermal cut-off switches might also be valuable if inadequate cooling is detected.

The reliability of the fans must be considered as well. If the forced convection by ventilation fails, for example, due to a blocked filter, and the result is an unacceptably high temperature in the system, the inside temperature should be monitored by a thermostatic switch and the system shutdown if necessary.

5.6   Recommendations for Thermal Management of Electronic Systems

The overall strategy with the thermal management of electronic systems should be to aim first for a sealed enclosure because we also need to keep shock protection, protection against ingress of foreign bodies or water, and electromagnetic compatibility in mind. If a sealed enclosure is impossible due to the large quantity of heat to be dissipated (see Table 5.11), ventilation from outside as free convection should be implemented by means of an open (perforated) enclosure. If this measure does not achieve the desired cooling, forced convection with fans or other functional elements should be deployed as a last resort.

Arrangement of heat sources in the system

The closer the heat source is to the top panel of an electronic system, the higher its temperature and the temperature of the top panel; however, the interior of the unit is then mostly cooler. Considering two identical heat dissipating enclosure areas, the maximum overtemperatures of the heat sources are lower for sealed enclosures with low overall height than those with a larger overall height.

Arrangement of printed circuit boards in an enclosure

Printed circuit boards are typically arranged in horizontal or vertical stacks. Boards should be arranged horizontally in small, compact systems for best thermal results, if the enclosure height/width ratio is less than 0.6. In all other scenarios, printed circuit boards should be installed vertically: There is better convection and temperatures are more evenly balanced. A clearance >30 mm is required between double-sided assembled printed circuit boards (single-sided boards ≥ 20 mm) so that unobstructed natural convection can more easily form; only 10–20 mm clearance is needed for forced convection.

Flow channels in the system

Vertical flow channels should be created if possible to facilitate the rising warm air. Head losses caused by flow resistances should be minimized. Channel constrictions and changes in the flow direction should be avoided.

Sealed enclosures

In sealed systems, heat is best removed from the heat sources to the enclosure by thermal conduction and by convection and radiation from the enclosure surface to ambient. Heat fluxes of up to 100 W/m^2 (i.e., 1 W per 10 cm× 10 cm enclosure surface area) can be dissipated through a sealed, bare metallic enclosure (mostly convection) for a temperature differential of 20 K, and up to 200 W/m^2 through painted surfaces (convection and radiation, all these values are approximations). Up to 180 W/m^2 could be dissipated through a 3-mm plastic enclosure with the same temperature differential.

Open enclosures

Airstream cross sections and panel cut-outs should be sufficiently large. Intake and exhaust vents should be located at the bottom and top of the enclosure. The reduction in the effective pressure at the intake and/or exhaust vents in the side panels should be compensated by larger panel cut-outs. The effective overall cross section of the airstream should be maximized. You should avoid inlet and exhaust vents of different sizes. Perforations should not exceed 25% of the enclosure surface (12.5% bottom and 12.5% top), as there are no additional benefits to be gained with larger openings. The enclosure base needs to be high enough off the floor (>30 mm) to ensure that the air can flow unrestricted into the enclosure through the air inlets in the enclosure bottom panel. It should be impossible to block the air vents in the top panel by placing objects on them.

Fans

When selecting a fan at the design stage, the required volumetric flow rate can often be estimated using only Eqs. (5.38a) and (5.38b); applications at high altitudes should consider the lower air mass delivered by the fan. The engineer typically has to resort to estimates and tests, such as measuring the static pressure at various flow rates, to obtain the required static pressure. The system should be designed so that more, or more powerful, fans can be used if required.

References

1. R. Reemsburg, *Thermal Design of Electronic Equipment (Electronics Handbook)*, CRC Press, 2000
2. *VDI Heat Atlas*, 2nd edition, Springer, 2010
3. G. N. Dulnjev, N. N. Tarnovski, *Teplovye rezimy elektronnoj apparatury*, Energija, 1971
4. www.fischerelektronik.de (January 2014)
5. B. Shelkpuk, "Heat Exchangers Cool Hot Plug-in PC Boards", *Electronics*, June 27, pp. 114-120, 1974
6. C. Markert, *Erwärmungsprobleme in elektronischen Geräten und ihre konstruktive Berücksichtigung*, Ph.D. Thesis, TU Dresden, Dresden, Germany, 1965

Chapter 6
Electromagnetic Compatibility (EMC)

An important topic in electronic systems design is electromagnetic compatibility (EMC), which concerns the unintentional generation, propagation and reception of electromagnetic energy. Every electronic system must meet EMC standards evidenced, for example, by the mandatory CE mark affixed to the system if sold inside the European Union or the EMC compliance mandated by the Federal Communications Commission (FCC) in the United States. The education and skill set of every engineer should therefore include a basic knowledge of EMC-related issues and their consideration in electronic system design (Sect. 6.1).

EMC is often the hidden culprit behind undesired circuit functionalities and unwanted signals. This is mostly due to the unintentional *coupling* of circuits that are either part of, only partially part of, or completely outside the system of interest. These couplings are dealt with in Sect. 6.2, and design options for their prevention, for example, by selecting appropriate *reference grounds*, are covered in Sect. 6.3.

While coupling deals with electrical circuits influencing each other with their fields, electronic assemblies and devices can also be disturbed by fields generated by external sources. Hence, one of the most important measures for assuring the EMC of systems is *shielding*. Section 6.4 introduces the principle of shielding, and discusses shielding against different types of fields.

A related discipline to EMC, *electrostatic discharge* (ESD), is covered in Sect. 6.5. The causes of electrostatic build-up and discharge are discussed as well as ESD protection measures.

Finally, we provide recommendations for good EMC practice in electronic systems design in Sect. 6.6.

© Springer International Publishing AG 2017
J. Lienig and H. Bruemmer, *Fundamentals of Electronic Systems Design*,
DOI 10.1007/978-3-319-55840-0_6

6.1   Introduction

The topic of "Electromagnetic Compatibility" (EMC) is concerned with the technical and legal aspects of the mutual interaction of electronic systems and their interaction with their surroundings through electromagnetic fields.

EMC guidelines, such as the "Blue Guide" on the implementation of EU product rules [1], stipulate that manufacturers and sellers of electronic systems are obliged to test these products for EMC. This ensures that equipment is developed to function properly in its electromagnetic surrounding, and that it does not interfere with other equipment. Two conditions must be met for compliance with these requirements. First, a system acting as an emitter should not emit unapproved disturbances (*interference*), and second, a system acting as a receiver should not be susceptible to external disturbances. The second aspect is a system's *immunity* to outside/external disturbances. Selective EMC measures should be put in place to maintain approved emission and immunity levels so that electronic systems can operate properly alongside one another.

The terms "emitter" and "receiver" do not refer to communication channels in this context, but more generally to all electronic systems, as many of them emit electromagnetic energy unintentionally as a byproduct of their operation. For example, every computer system emits energy due to its clock frequency; conversely, computers are also susceptible to disturbances: take, for example, communication problems with peripherals that are near disturbance sources such as an electric motor.

Disturbances, emitted by a *source of disturbance* ("culprit"), travel via a coupling path to the receiver, a *susceptible device* or *receptor* ("victim"). While disturbance sources and susceptible devices can easily be characterized by measuring their interference and immunity, insight into the physical fundamentals of electrical engineering is needed for identifying intermediate coupling mechanisms (see Sect. 6.2). And, knowledge of the coupling mechanisms is needed for selecting suitable countermeasures, such as shielding for sources as well as receptors (see Sect. 6.4), and thus for compliance with statutory emission and immunity levels.

6.2   Coupling Between System Components

Unwanted coupling may severely affect electronic systems. It occurs between the various components of an electronic system, such as cables, printed circuit boards (PCB) or chips. Disturbances couple from a component acting as an interference source to other components acting as receptors. Figure 6.1 depicts the four basic coupling mechanisms.

If the system dimensions L are much smaller than the wavelength λ of the electromagnetic waves at the frequencies f present in the system, the system is termed electrically small, i.e., $L \ll \lambda = c/f$. In this case, the wave that occurs can be ignored. The coupling in this situation can be classified as *conductive coupling* (also: *galvanic* or *common impedance coupling*), where there is a conductive

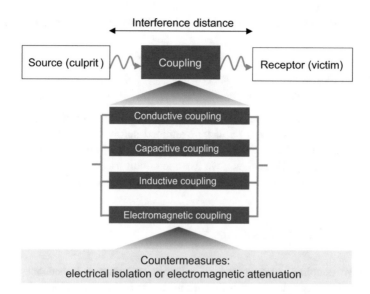

Fig. 6.1 Disturbances reach the victim (susceptible device) through four basic coupling mechanisms. Possible countermeasures are damping the source, hardening the receptor or attenuating the coupling path

connection between the source and the receptor; as *capacitive coupling*, where the coupling is caused by a varying electric field between the two adjacent conductors; and *inductive coupling* (also: *magnetic coupling*), where a varying magnetic field exists between two parallel conductors. This magnetic field is produced by varying currents, inducing an interfering voltage along the receiving conductor.

These three coupling modes are thus conductor bound. They can be modeled with simple equivalent circuit diagrams comprising discrete components, like resistors, capacitors and coupled inductances.

On the other hand, electromagnetic waves occur if a system is not electrically small, that is, the wavelength is on the same order of magnitude or smaller than the system dimensions: for example, the wavelength λ is only 30 cm for a 1 GHz frequency. The propagation of these waves can be either bound to conductors, such as cables or wave guides, or can be transmitted wirelessly, i.e., radiated. This phenomenon is called *electromagnetic coupling* (also: *radiative coupling*). Inductive and capacitive coupling effects cannot be separated in this scenario. Electromagnetic coupling should, therefore, be considered in electronically large systems with long cables, or systems at high frequencies.

All countermeasures should aim at identifying the disturbance sources; these sources should then be eliminated or their effects minimized. The coupling path between disturbance source and susceptible device, in particular, should be attenuated so that the susceptible device is not negatively affected by the disturbance. Hardening the receptor, i.e., raising its immunity limits, is also an option.

The following questions must be addressed to eliminate the disturbances and their effects:

– How do the disturbances arise?
– How do they penetrate a system?
– How do they impact the circuits?

To illustrate these issues, the four cases of coupling (conductive, capacitive, inductive and electromagnetic) will be examined below using two exemplary circuits. We assume that the two circuits are active, that is, they both possess power sources. To simplify the analysis, we assume circuit #1 (v_1) has the higher power rating and is the culprit; we shall examine its effect on the victim circuit #2 (v_2).

6.2.1 Conductive Coupling

Conductive coupling, also known as galvanic or common impedance coupling, occurs when the coupling path between the source and the receptor is formed by a direct electrical contact. Commonly, a current flows in a circuit ("culprit circuit") through an impedance that is also part of another circuit ("victim circuit"). The common impedance is often a common reference conductor or the common ground for different circuits on a printed circuit board. The current flowing in the culprit circuit produces a voltage drop across the common impedance, which superimposes itself on the signal in the victim circuit, and thus corrupting and possibly invalidating it. Conductive coupling leads to compatibility issues, such as the 50 Hz mains hum from many loads on the same supply line.

As shown in Fig. 6.2, currents i_1 and i_2 flow through the two circuits with common impedance Z_c, line resistance R_c and line inductance L_c. The potential difference v_c across the coupling impedance Z_c is proportional to the sum of two currents $i_1 + i_2$. The proportion of the external voltage based on the victim circuit (v_2) is due only to i_1. The voltage induced in the victim circuit is thus:

$$v_{c2} = R_c \cdot i_1 + L_c \cdot \frac{di_1}{dt}, \tag{6.1}$$

where v_{c2} is the interference voltage induced in circuit #2 in V, $\frac{di_1}{dt}$ is the rate of change of the current in circuit #1 in A/s, R_c is the resistance of the conductor common to both circuits in Ω, and L_c is the inductance of the conductor common to both circuits in H.

The induced voltage v_{c2} is in series with the input voltage v_2. Currents in electronic systems are generally not greater than a few amperes. One can often ignore the $R_c \cdot i_1$ component in Eq. (6.1) if the value of R_c is on the order of milliohms. Currents change rapidly in digital systems. Thus, relatively small currents through the inductance L_c produce a coupling effect if the currents are

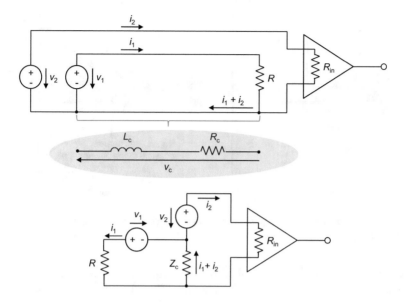

Fig. 6.2 Conductive coupling between two circuits (*top*) and equivalent circuit diagram (*bottom*)

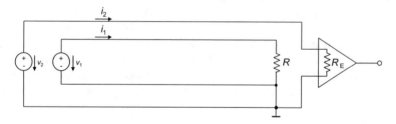

Fig. 6.3 Separate return lines to avoid common lines and thus conductive coupling

switched on or off rapidly. Coupling resistance R_c and coupling inductance L_c depend on frequency as well due to the skin effect (see Sect. 6.4.3).

Multiple options are available for reducing the coupling impedance. Optocouplers, coupling transformers, and the like, can be used to electrically isolate parts of circuits, for example. The use of separate ground and supply lines for loads drawing high currents (lamps, relays, power transistors, etc.) on the one hand, and logic circuits or analog amplifier circuits on the other, is encouraged. Reference and ground conductors should only be electrically connected at a single point. These two circuits are then electrically decoupled as they have no common lines or impedances (Fig. 6.3).

The single-point connection scheme is the preferred method for connecting different circuits. A star-shaped connection is typically selected, for example, to power analog and digital units separately (Fig. 6.4).

You should minimize coupling impedances if you cannot electrically decouple equipment with a single connection, which is often the case with digital electronics.

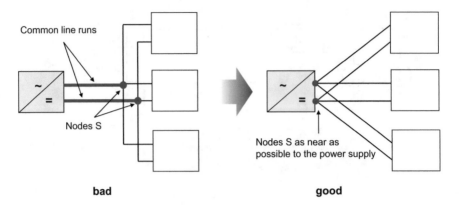

Fig. 6.4 Powering different components in a system via a single-point configuration avoids common line runs

The ohmic resistance can be limited by sufficiently large cross sections and low contact resistances at terminals and connectors. The line inductance should be kept to a minimum using techniques such as: using short lines in the common current path; using wide ribbon cables; routing signal and return lines close together (use thin film as a dielectric); and routing cables straight.

In practice, the engineer may have little influence on current transients caused by switching in the circuit. Voltage dips arising from short-term current spikes can however be mitigated by placing capacitors, such as *decoupling* or *energy-storage capacitors*, at suitable locations in the circuits. These capacitors are placed parallel to the power supply for every chip or piece of circuitry, and they will act as charge or energy storage devices in times of high current or high power consumption.

6.2.2   Capacitive Coupling

Capacitive coupling takes place between two circuits in which there is a potential difference, and thus an electric field, between the conductors in the circuits. If the voltage is time dependent, a displacement current will flow between the conductors and an additional line current of the same magnitude will flow in the conductors. The latter produces interference voltages in the impedances in the circuits, which are superimposed on the signal.

Capacitive coupling in the order of millivolts can occur, for example, in voltmeters if the sensor is located near a 120 V or 230 V power lead [2]. An electric field is produced in this scenario between the power conductor and the instrument conductor, which is almost at ground potential. This field produces displacement currents in the instrument conductor that corrupt measurements due to the field's stray capacitance between the conductors. Capacitive coupling is very much a function of the distance between culprit and victim; it occurs only at close range.

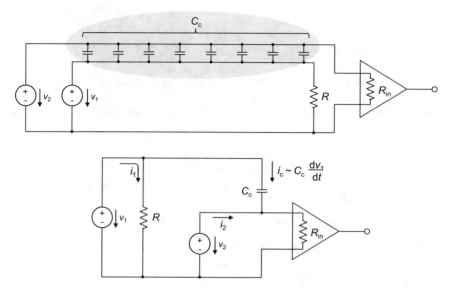

Fig. 6.5 Capacitive coupling between two circuits (*top*) and equivalent circuit diagram (*bottom*)

Capacitive currents flow from conductor to conductor via the capacitance C_c, which we have introduced in the circuit shown in Fig. 6.5. A partial current flows to the ohmic input resistor R_{in} in an amplifier and produces an ohmic disturbance voltage drop v_{c2}:

$$v_{c2} \sim i_c \cdot R_{in} \sim C_c \cdot \frac{dv_1}{dt} \cdot R_{in}, \qquad (6.2)$$

where v_{c2} denotes the interference voltage induced in circuit #2 at the conductor end in V, $\frac{dv_1}{dt}$ the voltage change in circuit #1 in V/s, C_c the line capacitance between the two circuits in F and R_{in} the terminating resistance at the end of conductor #2 or the amplifier input resistance in Ω.

The interference voltage increases with the line length and the frequency of the disturbance source, and as a result unshielded interconnects in LF and HF systems are exposed to large capacitive interference stemming from voltage changes in other circuits. Switching DC circuits produce interference voltages in the same way. These voltages drop with smaller terminating resistors at the start and end of the conductor.

The disturbance caused to other digital integrated circuits by capacitive coupling from digital signals is mitigated by lower signal amplitudes and greater signal rise and fall times. "Slow logic" will cause less disturbances than "fast logic" for the same signal level.

Interconnects with grounded shielding provide complete interference decoupling. The extra capacitance, which is 70–100 pF/m for coaxial cables, is a load for

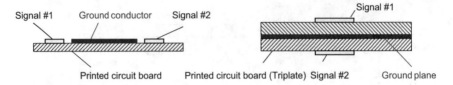

Fig. 6.6 Blocking the coupling capacitance with a shielding conductor (*left*) and shielding plane (on the *right*)

the drive circuit and reduces the logic speed in digital systems. Culprit and victim can also be isolated by placing shielding conductors (ground wires) between individual signal wires (Fig. 6.6 left), or by inserting shielding planes (ground planes) between circuit layers (Fig. 6.6 on the right).

Recommendations for reducing capacitive coupling are:

- Maximize distances between parallel signal lines.
- Decouple possible culprits and victims on PCB with shielding conductors.
- Do not use signal-carrying conductors in cable harnesses.
- Use point-to-point wiring for signal lines.
- Twist signal lines with additional ground lines.

As mentioned above, the steepness and amplitude of voltage changes within the culprit circuit should be kept as small as possible. Furthermore, the victim system should have minimal impedance. For further discussion on these modifications and their implementation, we refer to the literature, e.g., [3–5].

6.2.3 Inductive Coupling

Inductive coupling, also known as magnetic coupling, occurs between two current-carrying conductor loops. Current flowing in a conductor produces a magnetic field around the conductor. Inductive coupling occurs because a time-dependent current in the culprit circuit produces a time-dependent magnetic flux throughout the victim circuit inducing an interference voltage on the signal in this circuit (Fig. 6.7). The effects of the culprit current on the victim circuit are represented in an equivalent circuit schematic either by a mutual inductance (coupling inductance) or by an induced source voltage [2].

The voltage induced in the victim circuit is:

$$v_{c2} = -M_c \cdot \frac{di_1}{dt}, \tag{6.3}$$

$$\text{with} \quad M_c = k \cdot \sqrt{L_1 \cdot L_2}, \tag{6.4}$$

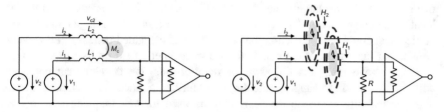

Fig. 6.7 Inductive coupling between two circuits by mutual inductance (*left*) and by a magnetic field (on the *right*). The depicted coupling actually acts over the entire length of the interconnects

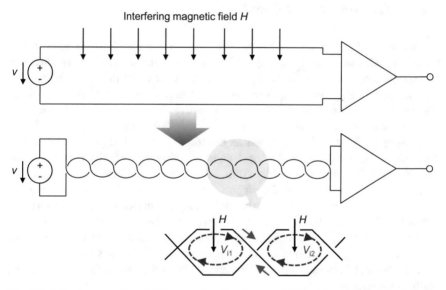

Fig. 6.8 Inductive coupling with extensive interference fields (*above*) and the induced interference voltages compensated by twisting (*bottom*). The voltage induction in the spaces between the twisted conductors alternates in its effect on the conductors (enlarged at *bottom*). The effective magnetic flux equals zero in the ideal case when the magnetic flux density is integrated across all component spaces

where v_{c2} is the interference voltage induced in circuit #2 in V; $\dfrac{\mathrm{d}i_1}{\mathrm{d}t}$ the change in current over time in circuit #1 in A/s; M_c the mutual inductance of the coupled conductor in H; L_1, L_2 the self-inductances of the conductors 1 and 2 in H, and k the coupling coefficient.

To reduce the inductive interference, the amplitude and steepness of the current changes in the disturbing system should be minimized. The coupling coefficient k can be reduced by increasing the distance between the coupled circuits. Inductances in conductor loops are at their lowest when signal and return lines are closest together. Hence, one should keep the surface area of pulsed conductor loops as small as possible. If the conductor is twisted, interference voltages from external magnetic fields are largely neutralized (Fig. 6.8). This is especially true for location-independent interference fields, such as in the vicinity of large transformers or electrical equipment. Twisting the conductor has the added benefit of effectively eliminating the effects of the conductor's own magnetic field on its surroundings.

If the above measures fall short of requirements, the development engineer will likely need to apply shielding to further reduce interference. Here, high-permeability magnetic materials may be used for low frequencies to divert magnetic fields from the victim (see Sect. 6.4.2).

6.2.4   Electromagnetic Coupling

In our initial analysis we dealt separately with circuit coupling via electric and magnetic fields in Sects. 6.2.2 and 6.2.3. For more detailed analysis we must recognize that both fields are, in fact, special cases of the electromagnetic field where the electric and magnetic components are linked according to Maxwell's equations. Electromagnetic coupling, also known as radiative coupling, exists if the victim is situated in a field where the electric and magnetic fields occur simultaneously.

If there exists at a given location an electric field that changes with time, then there also exists a magnetic field that changes with time, and which encompasses the electric field. This magnetic field, in turn, causes an electric field that changes with time, and which encompasses the magnetic field, and so on. The primary field can also be a magnetic field. The "disturbance" propagates in free space by this chain of electric and magnetic fields, thus creating an electromagnetic wave (Fig. 6.9).

The two different fields can be analyzed separately if the wavelength λ of the oscillations is large compared to the physical dimensions of the system of interest, and the distance between source and receiver. This is the case with slow processes, such as low-frequency circuits.

On the other hand, we can expect radiated and received energy from interconnects and devices in fast processes; in such scenarios signal circuits behave as antennas. As mentioned earlier, the wavelength λ is only 30 cm for a frequency of

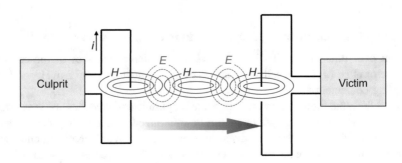

Fig. 6.9 Disturbance by electromagnetic radiation where the electric and magnetic fields occur simultaneously

1 GHz. Thus, the wavelength of the current and voltage is in the same order of magnitude as the physical dimensions of many electronic circuits, which causes a current and voltage wave to propagate in the circuit. The associated field is an electromagnetic radiation that disengages from the circuit and propagates through free space. Not only can such a circuit or such a system radiate electromagnetic waves ("culprit"), but it can also sense them and thus be disrupted by them as well ("victim", see Fig. 6.9).

Electromagnetic waves can be transmitted via interconnecting conductors as well as by radiation. In the former, the *wave travels between two conductors*. If the electric and magnetic fields emanating from this wire pair impinge on a second wire pair, electromagnetic interference will affect the second wire pair. The general case of coupling involves both capacitive and inductive interference occurring simultaneously.

In *interference by radiation*, on the other hand, an electromagnetic wave with electric field strength E and magnetic field strength H is the disturbance generated by a culprit (see Fig. 6.9). The fields E and H are orthogonal to each other. In the proximity of the culprit, i.e., in the near field at a distance $r \ll \lambda/(2\pi)$, either the electric field E (high voltages and low currents in the culprit) or the magnetic field H (high currents and low voltages in the culprit) dominates, depending on the type of culprit.

Shielding systems deployed to suppress conductor- *and* radiation-based inter-ference will be discussed in Sect. 6.4.6 (Shielding Electromagnetic Fields). Further countermeasures include techniques such as compact system construction, sym-metrical signal transport, and filtering of the power supply lines [4]. In addition, metal system enclosures should have only one entry point for power and signal lines to avoid HF currents over the enclosure.

6.3   Grounding Electronic Systems

6.3.1   Description of Reference Grounds

Input, output and source voltages in electronic systems are referenced with respect to a given *0 V potential*. This reference point is commonly called *ground*, even in reality there is no such thing as ground in electrical systems. If a system operates on battery, where is the ground? Nevertheless, we will use the term "ground" as the reference point in an electrical circuit from which voltages are measured.

Every circuit is ultimately referenced to a common point; the *system ground*. This low impedance conductor to a 0 V reference must be designed into the system right from the beginning. A good reference or grounding methodology avoids noise pickup and improves signal integrity by protecting against unwanted interference reception and radiated emissions.

Despite all the available options for conductive decoupling of different circuits in a system, there remains a connection in the form of a signal return path (ground). We shall examine what the influences on this reference potential are and how they affect their surroundings with an example and as a supplement to Sect. 6.2.1 on conductive coupling.

Figure 6.10 shows three different circuits Sl–S3 powered by the source voltage V_B via signal lines with line inductances L_1–L_6. The circuits (digital logic, amplifiers, power stages, etc.) are driven by control voltages v_1–v_3.

If there is a current spike di_3/dt due to a change in the control voltage v_3 in S3, the supply voltage V_{B2} for S2 is altered during the current spike by the value:

$$\Delta V_{B2} = (L_1 + L_2 + L_4 + L_5) \cdot \frac{di_3}{dt} + (R_1 + R_2 + R_4 + R_5) \cdot i. \qquad (6.5)$$

Furthermore, the base point of circuit S2 is moved by the voltage:

$$\Delta V_{O2} = (L_4 + L_5) \cdot \frac{di_3}{dt} + (R_4 + R_5) \cdot i \qquad (6.6)$$

with respect to the reference potential 0 V (minus terminal of V_B). The effective potentials at inputs S1 and S2 thus change as well, as the voltage drops at L_4 and $L_4 + L_5$ are added to input voltages v_1 and v_2 w.r.t. the ground potential. The combined input voltages may trigger unintended functions or outputs in S1 and S2. This unwanted voltage rise on the circuit ground is also known as *ground bounce*.

The change in supply voltage also typically causes malfunctions in the culprit circuit as well. The direction of the change in current (rising or falling edge) dictates whether the supply voltage is increased or reduced for the duration of the change in current.

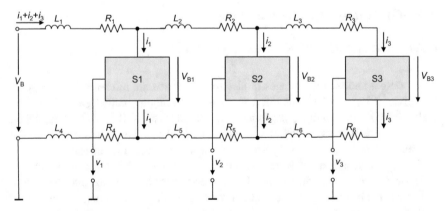

Fig. 6.10 Example of the disturbances to supply voltages by conductive coupling. Changes to the control voltage v_1–v_3 at one of the circuits S1–S3 alter the reference potential at the base points of the other circuits, which in turn impact their respective control voltages and can thus cause malfunctions

This explains how currents flowing in the reference or ground conductor can cause voltage drops, which in turn can cause system malfunctions due to changes in the effective input voltages. Conductor inductances have the greatest impact in the case of current transients and high-frequency alternating currents. High DC currents can also produce a significant voltage drop at the ohmic part of the conductor. This effect can be especially problematic in analog circuits.

Interference voltages induced in the reference or ground conductor should be avoided, as well as the disturbances caused by conductive coupling discussed above. The proposed countermeasures for inductive coupling (see Sect. 6.2.3) should therefore be used to mitigate interference effects in the reference or ground conductor.

Example 6.1 To gain better understanding, we illustrate the above effects of coupling disturbances on the reference conductor (ground) with a calculation using representative circuit values. Figure 6.10 is the equivalent circuit diagram for part of a commercial printed circuit board with circuits S1–S3. The length of the traces between the connector (plug) and S1 and between circuits S1–S3 is $l = 7$ cm, respectively. The trace inductance is approximately $L' = 10$ nH cm^{-1}. The resistance is $R' = 85$ mΩ m^{-1} = 0.85 mΩ cm^{-1} at 20 °C.

If a current of 20 mA flows through circuit S3 for 5 ns with a current slope of 4 mA/ns, an inductive voltage drop is produced at the reference conductor (connector to S3):

$$\Delta v_{O3(L)} = (l_1 + l_2 + l_3) \cdot L' \cdot \frac{di}{dt}$$

$$= (7\,\text{cm} + 7\,\text{cm} + 7\,\text{cm}) \cdot \frac{10 \cdot 10^{-9}\,\text{H}}{\text{cm}} \cdot \frac{20 \cdot 10^{-3}\,\text{A}}{5 \cdot 10^{-9}\,\text{s}}$$

$$= 0.84\,\text{V}.$$

The ohmic voltage drop:

$$\Delta v_{O3(R)} = (l_1 + l_2 + l_3) \cdot R' \cdot i$$

$$= (7\,\text{cm} + 7\,\text{cm} + 7\,\text{cm}) \cdot 0.85 \cdot 10^{-3} \frac{\Omega}{\text{cm}} \cdot 20 \cdot 10^{-3}\,\text{A}$$

$$= 0.36\,\text{mV}$$

is produced. An interference of this magnitude can generally be ignored in digital circuits; however, it will normally need to be taken into account in analog circuits.

6.3.2   Reference Systems Schemes (Grounding Systems)

There are two main reference systems for electronic circuits: a star-shaped single-point reference or ground (Fig. 6.11 left) and a reference or ground plane

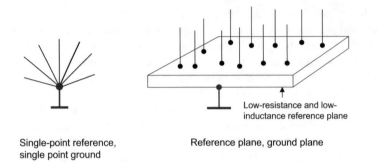

Single-point reference,   Reference plane, ground plane
single point ground

Fig. 6.11 Different grounding systems

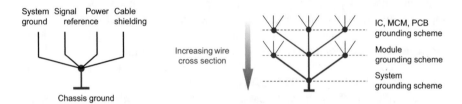

Fig. 6.12 Single-point grounding scheme (*left*) comprising system ground, signal and power reference and cable shielding ground, and hierarchical reference system (on the *right*)

(Fig. 6.11 right). The latter method of using a common and low impedance planar structure is sometimes also referred to as "multi-point ground".

In single-point grounding, the individual ground categories in a system all meet at a single point in an equipotential bonding busbar, called the *chassis* or *system ground* (Fig. 6.12 left). The main ground categories are:

- System ground (also: safety ground or chassis ground): to prevent an electric shock should a fault condition occur. It comprises all electrically conductive parts of modules and systems; the system ground for systems in protection class I (see Sect. 3.4.2) is connected to the protective earth PE, usually identified by a green-yellow color.
- Signal reference: all signal return conductors that provide a return path for intended signal flows.
- Power reference: all power supply return conductors, also known as the neutral conductor N, usually identified by a blue color.
- Cable shielding ground: all shields. The shield reduces electrical noise and interference with other devices.

A single-point ground terminal ensures that there are no potential differences between individual ground lines strapped to it. However, this type of ideal arrangement often becomes impractical to implement as it requires a massive amount of wiring. Hence, a hierarchical grounding scheme is recommended for large electronic systems, and is comprised of individual equipotential circuit planes (Fig. 6.12 on the right). The impedance of the ground lines should be gradually reduced from plane to plane in the direction of the system ground point to minimize any potential differences that may arise between the planes. This can be facilitated by increasing the cross sections of the ground lines in the direction of the ground terminal.

Low-impedance contacts with minimum contact resistance are recommended for ground conductors. This can be achieved by using planar contacts, avoiding oxide and coating layers by fitting locking washers, assuring contact force stability using spring washers, and preventing corrosion with suitable protective coatings. Electrochemical corrosion is another issue to be avoided by deploying connected partners in the electrochemical galvanic series with a low potential difference ($\Delta v \leq 0.25$ V).

A reference plane is especially beneficial at frequencies above 100 kHz, as then a circuit's ground potential is not only determined by the main ground point but also by other points in a circuit due to the effects of stray capacitances (Fig. 6.13 left). These stray capacitances can also cause unwanted loops with loop currents, which in effect neutralize the single-point grounding. It is only by implementing a reference plane as per Fig. 6.13 (on the right) that these stray capacitances can be fully isolated and a uniform reference potential achieved. Figure 6.14 shows an example of a reference plane on PCB.

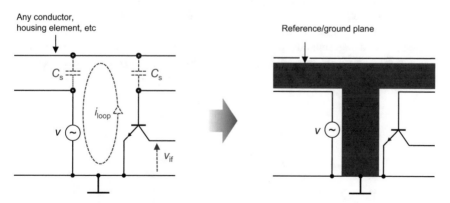

Fig. 6.13 The transition from a single-point ground (*left*) to a reference plane (on the *right*) avoids stray capacitances (C_s) and loop currents (i_{loop}) and reduces interference voltages (v_{if})

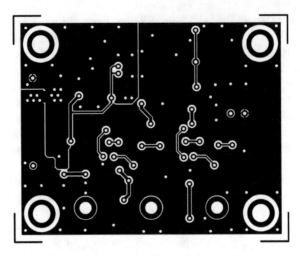

Fig. 6.14 Example of a reference plane characterized by internal wire tracks that are designed to "weaken" the reference plane as little as possible. A section of the circuit (*upper left* section) is ring fenced to decouple it from the remaining circuit

6.3.3   Return Conductor to the Reference Point for Digital Signals

When relays or contactors are controlled by digital outputs, the return signal from the load should not flow through the logic 0 V reference conductor. Figure 6.15 illustrates the need for separate return lines in such an arrangement.

In large systems, a dedicated power supply for each functional unit is recommended, as using a common power supply for many subsystems is the most common cause of disturbances. Figure 6.16 shows a schematic of a controller implemented using separate power supply units and ground lines. This type of configuration prevents mutual disturbances between functional groups occurring via the power supply. In addition, the interconnecting conductors between power

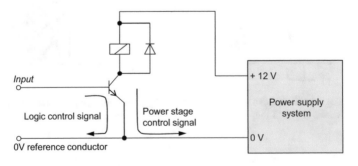

Fig. 6.15 Separate return line for the control signal of the power stage

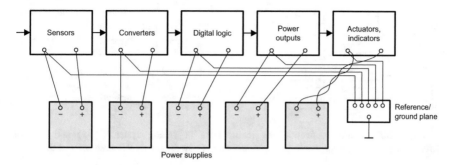

Fig. 6.16 Power supply configuration for an electronic system for preventing coupling via the power supplies

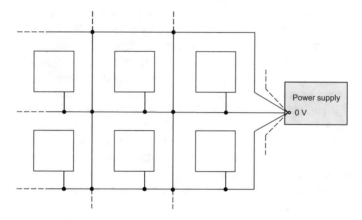

Fig. 6.17 Meshed topology for a low inductance ground conductor for digital logic

supplies and loads with large currents should be twisted to moderate magnetic field effects (see Fig. 6.8).

Inductance should be kept to a minimum in the digital logic due to the presence of transients (see Sect. 6.3.1). Very low inductances along with low wire resistances can be achieved with a meshed grounding scheme (and possibly a meshed power supply arrangement) as per Fig. 6.17. Ribbon cables have the lowest inductance. A meshed topology is a good approximation to the ideal planar structure.

6.3.4 Return Conductor to the Reference Point for Analog Signals

Input signals passing through input circuits for analog amplifiers often have very low values, with the result that disturbance coupling can cause major malfunctions.

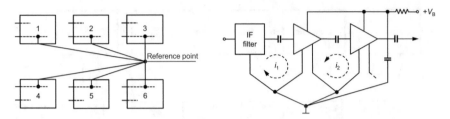

Fig. 6.18 Single-point connection for return lines in analog systems (*left*) and separate arrangement of ground terminals for input and output circuits for an RF amplifier (on the *right*)

Such sensitivity to interference is one of the reasons that the proper design of analog circuits is generally more challenging than for digital systems. Extensive experimental testing is often required for good outcomes.

As explained in Sect. 6.2.1, a single-point grounding scheme is used for analog input circuits to prevent conductive coupling (Fig. 6.18 left). If a number of circuits are connected at a single point, no current flows through the connecting reference conductor. The potentials of the individual reference points are thus all the same.

Grounding and shielding are critical for RF amplifiers, as parasitic oscillations[1] can be induced by coupling. We shall introduce only one standard measure against coupling by currents in ground conductors here. The different locations of the ground points for the input and output circuits in the RF amplifier as per Fig. 6.18 (on the right) ensure that the high-frequency input and output currents do not disturb each other via conductive coupling—which could lead to a feedback loop.

Shields (see Sect. 6.4) are treated as separate conductors and are also connected with the chassis or system reference. Shields must not carry any system currents.

6.3.5   Ground Loops

A power system should only be grounded at one place. Very large currents can flow in the earth. If power systems are grounded in many places, potential differences will arise between the different earth grounds. These "earth currents", which mostly originate from AC grid currents, are superimposed as mains hum on system signals in the event of a ground loop. The earth's resistance is parallel to the line resistance R_L of the return signal conductor (Fig. 6.19).

The same coupling issues arise if the chassis or rack is used as a return conductor for the power supply. If the reference potentials are unequal, a disturbance that is the same amplitude as the difference in reference potential is produced.

---

[1]Electronic devices can oscillate slightly in a circuit. The output of an operational amplifier can, for instance, oscillate at 100 kHz instead of outputting DC voltage. This phenomenon is due to internal parasitic capacitances that produce a phase shift that increases with increasing frequency to eventually become positive instead of negative feedback.

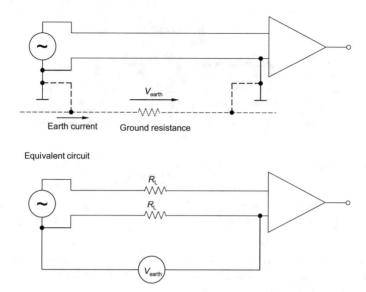

Fig. 6.19 Illustration of the effect of multiple grounds. A ground loop comprising a ground resistance parallel to the signal return conductor disturbs the effective signal of the circuit

Differential signal transmission (also: differential signaling) with two wires, which are not connected to the systems' ground conductors or are connected to a single point only, is a solution for these issues. The receiving circuit responds to the electrical difference between the two signals, rather than the difference between a single wire and ground (single-ended signaling). Symmetrical signal transmission ensures that the impact of disturbances affect both conductors (signals) identically. As the signal is transmitted with opposite polarity, the noise is subtracted away at the differential receiver. Variations in the reference potentials (grounds) of different systems are eliminated the same way.

6.4   Shielding from Fields

6.4.1   Shielding Fundamentals

Up to here, we have spoken about the effects of coupling on circuits, and how *their* fields interfere with each other. We want to move on now to the effects of external fields on modules and systems. In this regard, shielding is one of the key countermeasures to ensure EMC of systems (Fig. 6.20). An electrically conductive shield has the following key functions:

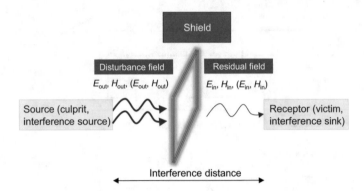

Fig. 6.20 Attenuating the disturbance by shielding (E_{out}, E_{in} external, internal electric field, H_{out}, H_{in} external, internal magnetic field, (E_{out}, H_{out}), (E_{in}, H_{in}) external, internal electromagnetic field)

- It blocks external electromagnetic fields from entering the system or attenuates them to such an extent that they cannot degrade system functionality (system immunity is thus assured),
- It prevents internal electromagnetic fields from leaking out of the system or it attenuates them to such an extent that the outside is not disturbed (an approved system emission level is thus assured).

Shields attenuate the electric and/or magnetic field intensity. We consider the case of an external electromagnetic field impinging upon a system that is protected by a shield. The ratio of the remaining field intensity H_{in} or E_{in} behind the shield ("inside") to the field intensity H_{out} or E_{out} of the external disturbance field ("outside") is often referred to as *shielding factor S*. This is a dimensionless quantity between 0 and 1; perfect shielding corresponds to a shielding factor of 0, while no shielding corresponds to a shielding factor of 1. The shielding factor for the magnetic field is given by:

$$S_{\mathrm{H}} = \frac{|H_{\mathrm{in}}|}{|H_{\mathrm{out}}|}, \tag{6.7a}$$

and for the electric field by:

$$S_{\mathrm{E}} = \frac{|E_{\mathrm{in}}|}{|E_{\mathrm{out}}|}. \tag{6.7b}$$

A more widely used metric, however, is the *shielding effectiveness SE*, which is defined as "the ratio of the signal received (from a transmitter) without the shield, to the signal received inside the shield; it represents the insertion loss when the shield is placed between the transmitting antenna and the receiving antenna" [6]. This power ratio is dimensionless and is expressed in logarithmic form (common logarithm, i.e., logarithm to base 10) in decibels (dB) for magnetic and electric fields as:

$$SE_H = 20 \cdot \log_{10} \frac{|H_{out}|}{|H_{in}|} \quad \text{and} \quad SE_E = 20 \cdot \log_{10} \frac{|E_{out}|}{|E_{in}|}. \tag{6.8}$$

Shielding factors for very good shields approach zero, and the shielding effectiveness should be at least 80 dB for very good shields. (The shielding effectiveness increases as the shielding factor decreases.) At 80 dB shielding effectiveness, the ratio of the disturbing (outside) to the residual (inside) field is 10,000 to 1, which means 99.99% of the disturbance is blocked (see Eq. 6.8).

Shields are used to block external (disturbance) fields. The job then for the engineer is to design a shielding with the most effective *shielding structure* and with suitable *shielding materials*.

It should be pointed out that the necessary bi-directional effect of shields (protecting the system from external disturbances from the outside world *and* protecting the system surroundings from interference from the electronic system) does not mean different design solutions are needed. Shielding electromagnetic radiation is the only exception here, where there are slight differences in the approach to shielding against emissions under near-field conditions, while protection against external disturbance fields is typically under far-field conditions (see Sect. 6.4.6).

The following descriptions of shielding apply to systems of any shape, size and "type", and to filters, coils, transformers or other items within a system, as well. Disturbance fields can be categorized as (Fig. 6.21):

- *Static fields.* These fields, also called constant fields, are constant over time and permanently distributed in free space. Static fields are the simplest kind of fields, as all terms involving differentiation with respect to time vanish. Examples of such fields are the electrostatic field of a parallel-plate, fully-charged capacitor and the magnetostatic field of a permanent magnet.
- *Magnetostatic fields.* These fields exist in the vicinity of currents that are constant over time and that flow through wires in one direction, and in the vicinity of permanent magnets. The magnetic field is stationary, i.e., its magnitude and direction remain the same at any given point.

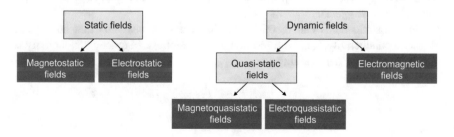

Fig. 6.21 Types of disturbance fields considered for shielding in this chapter

- *Electrostatic fields*. These fields exist in the surroundings of latent electrical charges. In other words, when two adjacent objects have different electrical charges, an electrostatic field exists between them. An electrostatic field also forms around any single object that is electrically charged with respect to its environment.
- *Dynamic fields*. These fields, also called variable fields, change over time and they can also change both spatially and over time. Dynamic fields are the most complex fields, since for them Maxwell's equations in their entirety must be satisfied.
- *Quasi-static fields*. These fields are a class of dynamic fields for which certain features can be analyzed as though the fields were static. Depending on whether the predominant static field is magnetic or electric, quasi-static fields are called *magnetoquasistatic fields* or *electroquasistatic fields*. In either case, both fields are time-dependent. The spatial field distribution is the same as with static fields.
- *Electromagnetic (wave) fields*. These fields are made up of interdependent electric and magnetic fields, which is the case when the fields are varying with time (dynamic fields). They are based on current and voltage propagation arising from a current and voltage wave propagation in a circuit (wavelength $\lambda \leq$ circuit dimension), that disengages from the circuit and radiates in free space. Electric and magnetic fields still behave like quasi-static in the near field, while the electromagnetic wave is dynamic and propagates over time in free space in the far field.

6.4.2   Shielding Magnetostatic Fields

Ferromagnetic materials are used to shield magnetostatic fields. This is because the magnetic reluctance of materials with high relative permeability μ_r is much lower than materials in the surroundings, such as air, for instance. The magnetic flux follows the path of least reluctance, as shown in Fig. 6.22. Hence, the flux is "drawn into" a highly permeable body, such as a shield made of a magnetic material. Magnetic field lines are always continuous and, in contrast to electric field lines, they do not terminate at the shield, but go through it. Thus, the disturbing field can be diverted around the object to be shielded through a magnetically conductive material (*bypass effect*).

Magnetostatic fields can therefore be shielded with materials that have good magnetic conductivity, i.e., with ferromagnetic materials. Magnetic conductivity is measured with the permeability μ:

$$\mu = \mu_0 \cdot \mu_r, \tag{6.9}$$

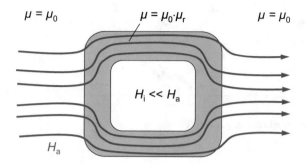

Fig. 6.22 Field distribution in a highly permeable shield with a static magnetic field (H_a external magnetic field strength, H_i interior magnetic field strength, μ_0 magnetic field constant, μ_r relative shield permeability). A shield of high permeability diverts a static magnetic field so that an object "inside" can be protected

with the magnetic field constant $\mu_0 = 4\pi \cdot 10^{-7}$ V s/(A m) $\approx 1.2566 \cdot 10^{-6}$ V s/(A m) and the relative permeability μ_r (ferromagnetic materials $\mu_r \gg 1$; vacuum $\mu_r = 1$).

Ferromagnetic materials have their own magnetic moments that are uniformly aligned in regions inside the material, called magnetic domains. Under the influence of an external magnetic field, these randomly oriented domains rotate in the direction of the field so that the permeability increases. If all the magnetic domains are oriented in the field direction, the magnetic flux is saturated. This field orientation, in the case of magnetically soft materials, is negated if the external field is removed. (Magnetically soft materials are easily magnetized and demagnetized.) Magnetically soft materials with high relative permeability μ_r are therefore used as magnetostatic shields (Table 6.1).

A shield must be able to allow magnetic field lines to pass through it without resistance and without saturating the material. A sphere—with its homogeneous shield material—is the most effective shape for a magnetostatic shield. The shield effectiveness of a sphere, a cylinder, and a cube are given in Fig. 6.23. An iron sphere with $\mu_r = 200$ and inner diameter $r_i = 50$ mm has a shield effectiveness of $a_s = 11$ dB for 1 mm shield thickness, and a shield effectiveness of $a_s = 28$ dB for 10 mm shield thickness according to Fig. 6.23. While the first value is inadequate for most shielding tasks, 28 dB equates to a reduction of the disturbance of more than 90%. Multiple shields made of a number of thin layers are a good choice for more demanding applications, as the shielding effect multiplies.

Table 6.1 Approximate relative permeability μ_r for different magnetically soft materials indicating their field distorting ability

Material	Supermalloy	Mu-metal	Permalloy	Nickel–iron	Carbon steel	Nickel
μ_r	100,000	25,000	4500	1000	200	100

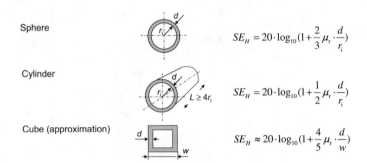

Sphere

$$SE_H = 20 \cdot \log_{10}(1 + \frac{2}{3}\mu_r \cdot \frac{d}{r_i})$$

Cylinder

$$SE_H = 20 \cdot \log_{10}(1 + \frac{1}{2}\mu_r \cdot \frac{d}{r_i})$$

Cube (approximation)

$$SE_H \approx 20 \cdot \log_{10}(1 + \frac{4}{5}\mu_r \cdot \frac{d}{w})$$

Fig. 6.23 Shielding effectiveness SE_H (in dB) for different shield geometries

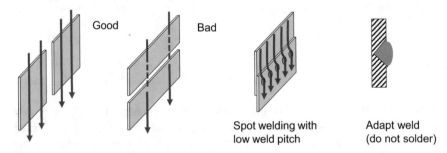

Fig. 6.24 Air gaps in the shield material should be avoided or they should be aligned with the disturbance to maintain a homogeneous field distribution (i.e., continuous field lines)

Bonded or press-fit connections should be used to join shield elements in order to avoid or reduce air gaps (Fig. 6.24). Only TIG welding with ferromagnetic materials should be used for bonded joints, as a soldered joint would be equivalent to a magnetic air gap because solder is non-magnetic. Distances between welds in spot-welded joints and between screws in shields screwed together should be as small as possible to minimize air gaps.

Magnetic shielding measures using magnetically soft metals are also deployed in addition to shielding magnetostatic fields (e.g., the Earth's magnetic field) for frequencies in the 50 Hz–20 kHz range. Layers of electrically-conductive metals like aluminum or copper are necessary for higher frequencies (see the next Sect. 6.4.3).

Although highly permeable materials provide effective magnetic shielding, they come with downsides: they are very sensitive to mechanical stresses and typically quite expensive.

6.4.3   Shielding Magnetoquasistatic Fields

A magnetoquasistatic field is a class of electromagnetic field in which a slowly alternating magnetic field is dominant. The effects occurring in the shield materials

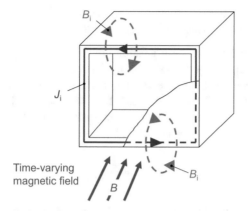

Fig. 6.25 The external, time-varying magnetic flux density B induces currents in the shield, which create a magnetic flux B_i in the opposite direction to the original field. The shielding effect is thus based on the resulting field attenuation in the interior and an additional field attenuation in the shield material at higher frequencies due to the skin effect

due to such slowly varying magnetic fields are caused by coupling between electrical and magnetic processes:

(1) magnetic flux changes are related to electric field strengths and thus voltages (Law of induction),[2]
(2) electrical current is related to a magnetic field as a consequence of an electric field strength (Ampère's circuital law).

An electrical eddy current field E_i is induced in the electrically conductive shield material due to the change in magnetic flux density B with time. This field drives a current density J_i that is enveloped by a magnetic flux density B_i that it generates. This magnetic flux density B_i inside the shield is in the opposite direction of the primary flux B as illustrated in Fig. 6.25. The disturbance field is thus attenuated inside the shielding enclosure.

The mechanisms described above produce the following two shielding effects:

(1) Attenuation of the disturbance field in the interior of the shielding enclosure by the induced eddy current in the shielding enclosure (induced "counter field" that opposes the applied magnetic field, i.e., a reduction in the internal disturbance field by superimposition of a secondary field).
(2) Field attenuation from the outside to the inside of the shielding plate by the *skin effect*. This describes the effect of the reduction in current from the outside to

---

[2]The direction of the eddy current generated is opposite to the right-hand rule according to the Law of induction. This is logical, because if a potential difference were produced in the direction of the right-hand rule as the flux increases, it would drive a current in a conductor that would increase the flux inside the loop and thus reinforce itself. In actual fact, the potential difference acts against its own cause (Lenz's law).

the inside of a metallic conductor as a function of the frequency and the electrical material constants of the conductor (permeability and conductivity).

The disturbance field can only penetrate the shield material to a certain depth depending on the frequency due to the skin effect described in (2). The *skin depth* δ is a critical parameter that characterizes this depth of penetration. It is the distance from the surface at which the disturbance field is attenuated by a factor of "$1/e$" (that is, approximately 37% of its value at the surface; Euler's number $e \approx 2.718$). The skin depth δ is defined for good conductors as:

$$\delta = \frac{1}{\sqrt{\pi \cdot f \cdot \mu \cdot \kappa}}, \qquad (6.10)$$

with frequency f, permeability μ ($\mu = \mu_0 \cdot \mu_r$), and electrical conductivity κ of the shield material.

Table 6.2 illustrates the theoretical skin depths δ for several different flat conductors. Ferromagnetic conductors are needed for low frequencies because they facilitate eddy currents. Field attenuation increases considerably at high frequencies; current flows only in a very thin layer near the surface—this is why this phenomenon is called the skin effect.

Low skin depth δ and thus high shielding effect is achieved with

- a high frequency disturbance field,
- a highly permeable shield material, and
- a shield material with high electrical conductivity.

The shielding effectiveness SE_H is calculated as:

$$SE_H = 20 \cdot \log_{10}\left(e^{d/\delta}\right) \approx 8.7\frac{d}{\delta}, \qquad (6.11)$$

with shield thickness d and skin depth δ as per Eq. (6.10). A desired attenuation of the disturbance field can thus be achieved to below 1% (SE_H = 43.4 dB) with a shield thickness five times greater than the skin depth. (This is where the well-known design rule $d \geq 5\delta$ for shield thickness comes from.)

Shielding enclosures must be designed to allow eddy currents to flow freely (Fig. 6.26) as shield effectiveness at higher frequencies is determined by the

Table 6.2 Skin depth δ for different flat conductors and frequencies to illustrate the skin effect

Material	δ in mm				
	50 Hz	800 Hz	1 MHz	100 MHz	10 GHz
Copper	9.6	2.4	0.067	$6.7 \cdot 10^{-3}$	$6.7 \cdot 10^{-4}$
Aluminum	13.3	3.3	0.94	$9.4 \cdot 10^{-3}$	$9.4 \cdot 10^{-4}$
Iron (μ_r = 300)	1.5	0.38	0.011	$1.1 \cdot 10^{-3}$	$1.1 \cdot 10^{-4}$
Mu-metal (μ_r = 25,000)	0.333	0.084	$2.36 \cdot 10^{-3}$	$2.36 \cdot 10^{-3}$	$2.36 \cdot 10^{-5}$

Fig. 6.26 Ensuring
undisturbed eddy currents

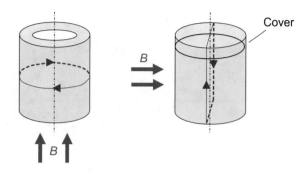

magnitude of the eddy currents, i.e., to which level the impedance of the shield can
be reduced. Joints and seams should be normal to the direction of the disturbance
field. Metallic joints and seams should be smooth when they run in the same
direction as the disturbance field. The resistances of joints and seams in the
shielding enclosure, which may be unavoidable for assembly and operational reasons—a removable cover might be needed for the enclosure, for example—should
be minimized as the direction of the disturbance field is typically unknown.
Conductive contacts or seals are recommended for such a scenario.

In addition to eddy currents, the bypass effect mentioned above also contributes
to shielding in the case of low-frequency time-varying magnetic fields. As with
high-frequency magnetic and electromagnetic fields, the magnetic fields caused by
eddy currents attenuate these disturbance fields so much that highly permeable
materials are not needed. Shields with high electrical conductivity, such as copper
and aluminum, that are connected to the reference potential, should be used here
(see Sect. 6.4.6).

6.4.4 Shielding Electrostatic Fields

In Sect. 6.2.2 we described how capacitive interference can be suppressed with
grounded shields. In response to an external electric field, charge on an electric field
shield redistributes itself; at the same time it is itself a source of an electric field, as
well. As illustrated in Fig. 6.27, induced charges on the outer surface of a conductive object will shield against electric fields as there is no surface charge along
the inner surface. Thus, an electrically conductive object in an electrostatic field E_{out}
is an ideal electrical shield since there is no field inside the body (electric field
strength $E_{in} = 0$). This is also known as a *Faraday shield*.

The charges are redistributed in the shield material by the forces
$F = E_{out} \cdot Q$ acting on the mobile electric charges Q by the external field E_{out}. The
charge redistribution is terminated when the field of the displaced charges and the
external disturbance field cancel each other out in the shield interior. Thus, there are
no potential differences on the inner surface of the shield, and $E_{in} = 0$ applies for

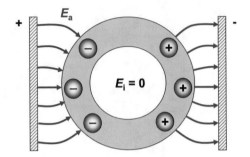

Fig. 6.27 The Faraday shield effect is based on the superimposition of the external electrical disturbance field on the field produced by the charges induced on the surface. There is then no electric field in the system interior, as charges are redistributed on the inside shield surface until there is no more potential difference on this inner surface

internal field strength. Hence, the shielding effectiveness of a smooth electrically conductive shield tends to infinity ($SE_E = \infty$).

These deliberations on the above mechanisms lead us to conclude that the shield does not have to be made of a continuous material, but can also be a conductive metallic mesh. A well-known example is the *Faraday cage*.

We find similar conditions for electric fields inside electronic systems. For example, a live conductor is surrounded by a metallic shield in Fig. 6.28 (left). The field lines of the positive charge of the core conductor terminate at negative charges on the inner shield surface. New field lines emanate from the positive charges induced on the outer surface to the surroundings and end up at appropriate charges on the ground or the surrounding walls.

On the other hand, if the shield is grounded, the field lines of the positive conductor also terminate at negative charges on the inner surface of the shield (Fig. 6.28 on the right), while the external positive charges flow to ground. Conductor and inner surface of the shielding form a closed system from which no field lines emanate to the surroundings if the shield is a continuous entity. Coaxial cables are based on this design principle.

Applying AC voltage, the internal field is continuously reversed (see the next Sect. 6.4.5). Similarly, the induced charges on the outer shield surface are exchanged via the ground conductor. Obviously, the impedance of shield and conductors should be minimized. Conductor cross sections should be large enough and the line length to the reference conductors should be minimized, as well.

All side panels in an electrically conductive shielding enclosure must be joined at multiple points, at the least, for the required equipotential conditions, i.e., electrical potential equalization. A minimum of three joints are needed. This type of shielded enclosure typically requires no ground conductor (Fig. 6.29).

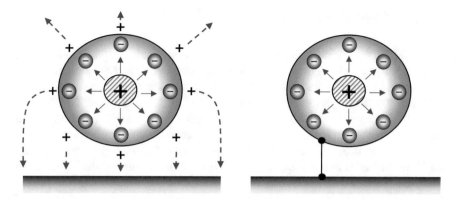

Fig. 6.28 Electric field of a cable with inner and outer conductors. The electric field of the "floating shell" on the *left* emanates from the positive charge on the surface; this configuration does not function as a shield. Shielding can be achieved by grounding the outer conductor and by the resulting discharge of the external positive charges (on the *right*)

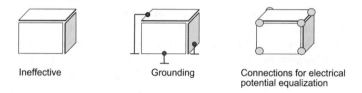

| Ineffective | Grounding | Connections for electrical potential equalization |

Fig. 6.29 Side panels that are not joined together (*left*) adopt the potential of the active field at this location and provide no shielding. Here, shielding can be provided only by grounding the panels (*center*). Equipotential connections are required if there is no ground (Faraday cage, on the *right*)

A non-conductive (dielectric) object can also carry an electric flux Ψ due to its "dielectric conductivity", i.e., a relative permittivity $\varepsilon_r \gg 1$. This bypass effect with respect to the surroundings with $\varepsilon_r = 1$ yields $E_{in} < E_{out}$. Thick-walled masonry or plastics are thus suitable for shielding electrostatic fields as well.

6.4.5   Shielding Electroquasistatic Fields

An electroquasistatic field is a class of electromagnetic field in which a slowly alternating electric field is dominant (approximately: $f < 100$ kHz).

With regard to the shielding of electroquasistatic fields there is a phase shift in the charge distribution on the shield surface as the frequency increases, yielding a finite shielding effectiveness. As this effect only becomes noticeable at very high frequencies, we can make the engineering assumption of an infinitely large shielding effectiveness for electroquasistatic fields as well. This allows the application of the same shielding design rules as for electrostatic fields (Sect. 6.4.4).

6.4.6   Shielding Electromagnetic Fields

A useful limit for transitioning from the prior discussed static considerations to high frequency behavior is defined by the so-called *λ/10 rule* [4]: If the structure to be investigated is smaller than 1/10 of the smallest wavelength to be considered, it is sufficient to use the quasi-static approaches and considerations. Otherwise, i.e., the structure is larger than 1/10 of the wavelength of the highest frequency, high frequency behavior, where inductive and capacitive coupling effects cannot be separated, has to be assumed. Hence, using a clock frequency of approximately 83 MHz (or higher) on a computer main board with the dimensions 20×30 cm (diagonal size: 36 cm) requires the consideration of electromagnetic fields.[3] Electromagnetic shielding should, therefore, be considered in large systems or at high frequencies.

Electromagnetic fields or waves can spread by means of conductors (*conductor-borne*) and in free space (*interference from radiation*).

The *conductor-borne* wave travels between two conductors. If the electric and magnetic fields from this conductor pair disturb another conductor pair, the latter will be subjected to electromagnetic emission (EMI) from the former. This simultaneously acting capacitive and inductive interference is the general case of coupling.

When disturbance coupling occurs between interconnects, an EMI is always expected if the conductor length is on the same order of magnitude as the wavelength λ_s of the frequencies f_s of the disturbance. It has been found in practice that in digital electronics with an operating frequency of 100 MHz (the maximum disturbance frequency will be $f_s = 400$ MHz in this case), the conductor length is $l_s \approx 8$ cm ($l_s \approx \lambda_s/10$) at the start of EMI.

Shields are used to direct conductor-borne EMI so that it does not disturb susceptible components. In this regard, *coaxial cables* are a common solution: the core conductor carries the high-frequency signal and the exterior jacket the return signal. Transmitter and receiver, in contrast to capacitive shielding (see Sect. 6.4.4), must connect their reference potential to the outer conductor of the coaxial cable ("all-round" contact), so that the return signal can flow through it (Fig. 6.30).

The *transfer impedance* Z_t is a measure of the shielding effect for conductor-borne EMI. A schematic of a coaxial cable is shown in Fig. 6.31. If a parasitic current I_s flows through the (shorter) outer conductor (length $L < \lambda_s/20$; λ_s wavelength of the disturbance voltage), the voltage across the open-circuit input terminals of the conductor is V_2 (the conductor is shorted at the other end). We can express the transfer impedance per unit length Z_t as:

$$Z_t = \frac{V_2}{I_s \cdot L}. \qquad (6.12)$$

---

[3]The wavelength λ of $f = 83$ MHz is approximately 3.6 m; λ (m) $= c$ (m/s) $/ f$ (Hz), simplified λ (m) $\approx 300 / f$ (MHz).

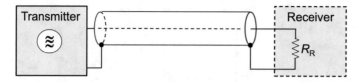

Fig. 6.30 Shielding conductor-borne electromagnetic waves through a coaxial cable, where transmitter and receiver are connected to their reference potential at the outer conductor of the coaxial cable. When the two opposite currents on the inner and outer conductors are balanced, the field emission from the coaxial cable is zero

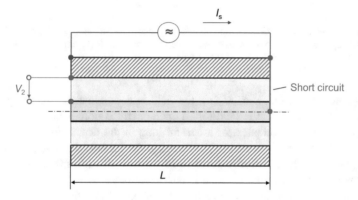

Fig. 6.31 Cross section of a coaxial cable with the quantities needed to determine the transfer impedance per unit length Z_t

The standard unit for Z_t is mΩ/m. The external parasitic current I_s produces a disturbance voltage V_2 in the internal system by conductive coupling with the outer conductor. This disturbance voltage is small if the transfer impedance is small. An outer conductor with a high shielding effect has a low transfer impedance. We illustrate in Fig. 6.32 several important relationships regarding transfer impedance. The transfer impedance Z_t is approximately the same as the DC resistance R of the outer conductor with DC and low frequencies. However, as the frequency rises, the discrepancy between the transfer impedance and the DC resistance increases considerably, whereby the value of the impedance and dependency upon the frequency depend on the design of the outer conductor.

Current flow via shields should be avoided to prevent conductive coupling issues. The coaxial cable described above is an exception, as the outer conductor carries the return signal for the inside conductor.

In the case of *interference from radiation*, the culprit device generates electromagnetic waves with electric field strength E and magnetic field strength H (see Fig. 6.9). The field and wave propagation characteristics depend on the distance r between observer and emitter.

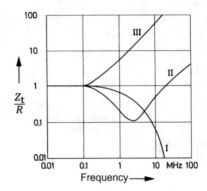

Fig. 6.32 Frequency dependency of the transfer impedance for shielded conductors. The transfer impedance Z_t of the shield is normalized w.r.t. the DC resistance R (I homogeneous, RF-shielded conduit, II wire mesh, III helically wrapped tape or film)

In the near field, i.e., $r \ll \lambda/(2\pi)$, the field is determined by the design of the radiating circuit. The electric field for an open circuit is dominant in the near field. The magnetic field dominates for a current loop in the near field due to coupling with the current. A body is said to be located in the far field when it is at a distance $r \gg \lambda/(2\pi)$ from the emitter. Regardless of the type of radiating circuit, the electromagnetic radiation in the far field is non-stationary.

In order to calculate shield attenuations the Maxwell equations must be solved—a task which is quite challenging mathematically. The *Schelkunoff impedance concept* is a much simpler approach. This concept considers the impedance conditions in front of, behind, and in a shield. The mechanisms at play such as reflection at impedance boundaries, absorption in the shield material and transmission are illustrated in Fig. 6.33. The shielding effect is based on

- the absorption loss A due to absorption in the shield material,
- the reflection loss R based on reflections at impedance boundaries Z_{out}/Z_{screen} and Z_{screen}/Z_{in},
- a re-flection coefficient B, a reduction in attenuation (negative value) due to multiple shield penetrations caused by multiple internal reflections, where $B \approx 0$ for $A > 10$ dB applies [2].

The shielding effectiveness SE can thus be expressed as:

$$SE = A + R + B. \tag{6.13}$$

The Schelkunoff impedance concept and its application using Eq. (6.13) typically simplifies the calculations for the shielding effectiveness of electromagnetic radiation to the sum of *absorption loss* and *reflection loss*.

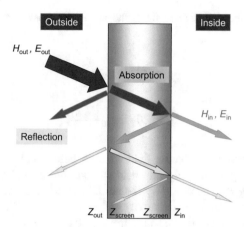

Fig. 6.33 The parameters in the Schelkunoff impedance concept for determining the shielding effectiveness based on absorption and external and internal reflections

The *absorption loss A* arises from the energy loss in the shield resulting from the eddy currents generated by the radiation. The absorption loss is directly proportional to the shield thickness and proportional to the square root of the frequency, the radiation barrier's conductivity and permeability. A plot of the absorption loss as a function of the frequency is given Fig. 6.34 for different metal barriers. The reader is referred to the literature, e.g. [5, 7], for more detailed calculations.

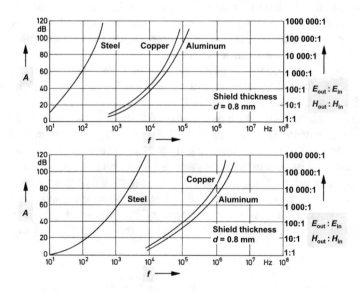

Fig. 6.34 Absorption loss A for shields of various metals and barrier thicknesses as function of the frequency f (E_{out}, E_{in} electric field strength before and after the shield, H_{out}, H_{in} magnetic field strength before and after the shield)

Depending on the disturbance field, the absorption loss in the shield in dB is defined as:

$$A_E = 20 \cdot \log_{10} \frac{|E_{out}|}{|E_{in}|}, \tag{6.14a}$$

$$A_H = 20 \cdot \log_{10} \frac{|H_{out}|}{|H_{in}|}, \tag{6.14b}$$

where E_{out}, E_{in} denote the electric field strength before and after the shield, and H_{out}, H_{in} the corresponding magnetic field strength before and after the shield.

The *reflection loss R* is caused by the partial reflection of the incident wave on the shield. It depends on the conductivity and permeability of the metal barrier, the frequency, and the distance of the shield from the culprit device. These are complex issues and are beyond the scope of this book (see [4, 7] for further details).

Good electrical conductivity is a key property of a material for an RF-shielded barrier, similar to an alternating magnetic field. The enclosure should be completely sealed to allow eddy currents unrestricted passage. Joints should be electrically bonded along their entire length by means of soldering, for instance, or alternatively sealed by twin-contact spring plates or metallic rope seals.

A metallic enclosure with no openings is an excellent shield; openings significantly reduce the shielding effectiveness. Due to the physics of field penetration, many small openings are preferred to a few large ones; round holes are better than square ones; and square cut-outs are better than rectangular ones. An opening is said to be small if its diameter is small compared to the wavelength of the incident radiation. Radiation emanating from a circular hole produces a field strength proportional to the cube of the diameter of the hole. Hence, the field strength increases by a factor of eight if the diameter is doubled.

The largest perforation determines the scale of the reduction in the shielding effect. The shielding effectiveness will be about 0 dB (i.e., no shielding) for a perforation length/diameter of 30 m / f (in MHz) ("$\lambda/10$ rule"). Hence, the shielding effectiveness reduces to 0 dB for an opening (slot, crack in the cover, and the like) of 30 cm for frequencies greater than 100 MHz. We can thus determine the maximum slot length for a known operating frequency; for example, slots longer than 10 cm should not be permitted for frequencies on the order of 300 MHz ($\lambda \approx 1$ m).

If large holes are required, they should be implemented as honeycomb metallic structures as they allow ventilation while attenuating electromagnetic fields below the cutoff frequency[4] of each of the "waveguides" in the structure [7]. The shielding effectiveness *SE* of a honeycomb mesh depends on its depth L and diameter d.

---

[4]The cutoff frequency is the frequency beyond which a waveguide no longer contains EMI. For a rectangular waveguide, the cutoff frequency (in MHz) is $f_c = c / 2d$ with c the speed of light (in m/s) and d the largest cross section of the waveguide (in m).

Fig. 6.35 Honeycomb windows provide RF shielding that is proportional to the depth L and inversely proportional to the diameter d

The depth should be maximized and the diameter minimized because $SE \sim L/d$ (Fig. 6.35). For example, for honeycomb filters where $L > 4d$, the absorption loss is greater than 100 dB [7].

6.5 Electrostatic Discharge (ESD)

Electrostatic discharge (ESD) is the sudden flow of electricity between two electrically charged objects. More precisely, it is an electrical discharge from an electrically isolated material with a high potential difference that causes a very short and high electrical current pulse. An ESD event can be felt at about 3 kV and higher. However, only a fraction of this voltage is needed to damage or destroy electronic devices. ESD issues therefore must be considered in electronic system development to prevent system failures.

6.5.1 Causes of ESD

One of the most common causes of ESD is static electricity, which readers may be familiar with in cold, dry weather when combing their hair, or the tiny shocks that may result from walking across a carpet and then touching a piece of metal. The underlying potential difference is primarily caused by the *triboelectric effect* (also known as *charging by friction*). Here, two materials with different energy levels for electrons (Fermi levels, i.e., the energy of the highest energy electrons at zero temperature) are in frictional contact; i.e., they are subjected to frictional forces between one another. Electrons tend to move from the material with high electron energy levels to the material with lower energy level. Hence, charges are transferred between the bodies while they are in contact; the electron charges are unbalanced when the bodies are separated again. Because the body's surface is now electrically charged, either negatively or positively, any contact with an uncharged object, such as a grounded body, may cause an electrical discharge of the built-up static electricity (Fig. 6.36 left).

An ESD event can also be triggered by a process known as *induction*: for example, electronic components can be polarized if they are in the vicinity of an electric field. Such a field can come from equipment or feed belts that have been

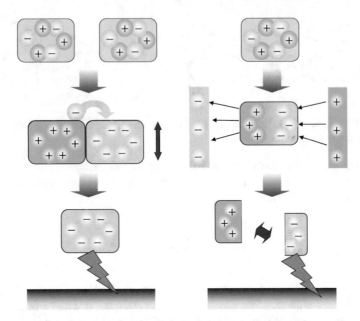

Fig. 6.36 The effect of charging by friction (*left*) and induction (on the *right*) can lead to electrostatic charge buildup followed by subsequent discharge. Electrons are transferred in the scenario on the *left* as a result of friction between two objects of different Fermi levels, the potential difference in the scenario on the *right* is induced by an electric field followed by the separation of an object

previously charged with an electrostatic charge. Even though the net electrostatic charge of the object has not changed, it now has regions of excess positive and negative charges. When components are separated from the object, for example, components from a feed belt, a potential difference is produced with the risk of ESD (Fig. 6.36 right).

Electrostatic discharge can damage microelectronic devices, and consequently cause the device or the entire system to fail. Components can be severely damaged or partially degraded by sparkovers, which can lead to immediate failure or subsequent failure in the field. Integrated circuits are particularly susceptible to ESD, and it is one of the main causes of catastrophic damage to chips, as their typical cutoff voltages are between 5 and 30 V. Measures need to be put in place to prevent even small amounts of static build-up.

6.5.2 ESD-Suppression Measures

Electrostatic discharge is prevented by (1) safely discharging unavoidable parasitic static build-up and (2) minimizing the build-up of electrostatic charge in the vicinity of electronic devices [8].

While integrated protection circuitry ensures that static build-up is discharged within electronic components, such as integrated circuits, an electrostatic discharge protected area (EPA) comprises a range of protection measures for manufacturing and handling these components. EPA-classified manufacturing areas are designed to prevent static charge build-up. This is achieved by electrically conductive and grounded workbenches, the use of conductive filaments on garments worn by assembly workers, conductive wrist and foot straps and floor covering, ionized air and the grounding of all devices. The air humidity should be at least 50% as well.

Packaging for ESD sensitive components should be made of electrically conductive and electrostatically dissipative plastics; it acts as a Faraday cage to protect the contents against ESD. Conductive or metalized films, padding and foams are typically used for this ESD protective packaging. We recommend that external pins for devices are connected with a shorting jumper for handling and transportation.

6.6   Recommendations for EMC-Compliant Systems Design

6.6.1   Key Steps in System Development

The mechanical design, the layout of the PCB, and use of the grounding concept have a big influence on the EMC of a system. Debugging EMC problems at a later stage by shielding and filtering, for example, involves extensive effort and should be avoided by proper design. The key steps in an EMC-compliant electronic system development are pictured in Fig. 6.37.

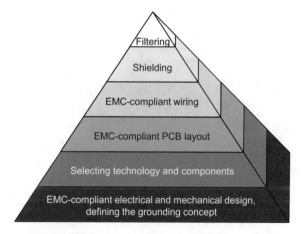

Fig. 6.37 Key steps in EMC-compliant electronic system development [4, 9]. The size of a pyramid layer is an indication of the importance and the work attached to each measure; the order of their execution is indicated on the vertical axis (from *bottom* to *top*)

6.6.2   Designing Printed Circuit Boards and Shielding

The following 25 rules for the EMC-compliant design of PCB and shielding enclosures have been proven effective and are recommended for practical designs. We have cited these rules and Figs. 6.38, 6.39, 6.40, 6.41, 6.42, 6.43 and 6.44 from [4] with the kind permission of the authors, and adapted them as needed for the requirements of electronic systems design.

Rule 1: Physically group functional units together (and thus separate them from other groups) on a printed circuit board or in a system (Fig. 6.38).

Rule 2: Power supply lines should be filtered at the input (connector) on a printed circuit board.

Rule 3: Different types of circuits (analog, digital, power supply) should be grounded separately. Use multi-layer PCB with one or more ground planes for 10 MHz clocks or higher. It is wise to have ground tracks in place to the left and right of RF signal lines if you cannot use separate ground planes for cost reasons.

Rule 4: Keep the areas enclosed by loops at the power supply to a minimum. Small loops cap radiation and enhance immunity (Fig. 6.39).

Rule 5: Avoid current loops. Circuits require signal and return conductors; run signal and return conductors close to each other on a printed circuit board (Fig. 6.40). EMI and interference coupling are approximately proportional to the loop area.

Rule 6: Implement defined return paths for currents. Current always flows through the path of lowest impedance. A return current path always establishes itself in the proximity of the (forward) signal path in the case of RF signals. Defined return current paths are needed to mitigate radiation and immunity issues. The return signal automatically takes the path of lowest impedance on a printed circuit board with ground plane; coupling between two circuits via the impedance of the return current path is generally low (Fig. 6.41 left). Run the return line near the signal line

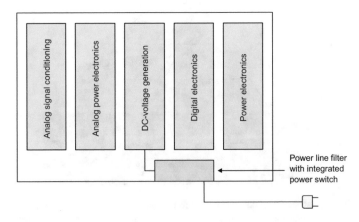

Fig. 6.38 Modular system layout

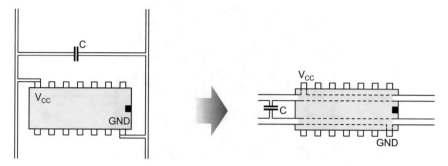

Fig. 6.39 Optimizing the power supply for a circuit (poor design on *left*) by redesigning and downsizing loops (improved design on *right*)

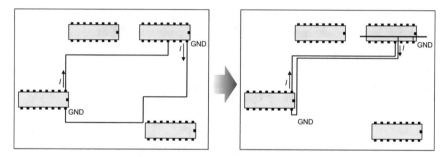

Fig. 6.40 Minimizing the loop area in a circuit to reduce interference emissions and inductive and capacitive coupling

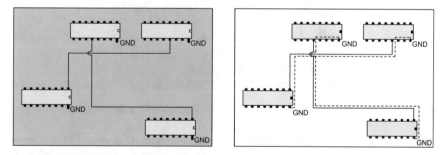

Fig. 6.41 The return signal takes the path of lowest impedance on a ground plane (*left*). Handling signals crossing each other on a printed circuit board with no ground plane (on the *right*) where the return conductor has to be placed near the forward current path and signal crossings have to be treated identically between forward and return conductor

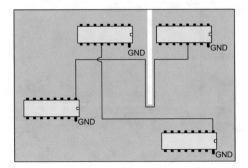

Fig. 6.42 Routing the (forward) current track around a break in the ground plane

Fig. 6.43 Preventing capacitive coupling by inserting a ground trace between two signal traces. This trace arrangement reduces the capacitance from 2.5 to 0.35 pF for a 10 cm parallel trace of width $w = 0.5$ mm and thickness $t = 35$ μm, and a substrate thickness $h = 1.8$ mm

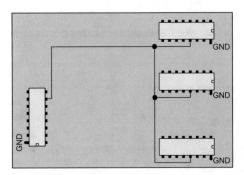

Fig. 6.44 Routing a clock signal to a number of devices by splitting the trace near the devices

on PCB with no ground plane, ideally taking an identical route on a different layer. Implement the return line in the same way as the (forward) signal line when both lines cross (Fig. 6.41 on the right).

Rule 7: Place a bypass or energy-storage capacitor (ceramic, 0.001–1 μF) in the vicinity of every load to keep current spikes local and to prevent currents from flowing through long conductor loops (Rules 4 and 5). Choose the resonance frequency of the capacitor so that it is higher than the maximum signal frequency.

Rule 8: If there is a break in the ground plane in a printed circuit board, route the (forward) current trace around the break (Fig. 6.42).

Rule 9: If you want to provide capacitive decoupling between two signal traces on a board, introduce a grounded trace between the two traces (Fig. 6.43).

Rule 10: Route lines with rapidly switching signals, that is, with high current spikes (di/dt) or high voltage spikes (du/dt), away from "sensitive" lines, such as analog inputs.

Rule 11: We recommend keeping clock lines short, and laying them square to signal lines. If you want to take clock pulses from the printed circuit board, place the clock generator as near as possible to the board connector (plug).

Rule 12: Decouple aggressive circuit components with RC, LC or RLC filters to stop disturbances from impacting the rest of the circuit. Do not forget to consider the resonance frequency.

Rule 13: Use terminated transmission lines based on their characteristic impedance (using what is known as the microstrip technique) in order to connect components on PCB for RF applications (clock > 100 MHz). Make every effort to avoid reflection and transition points.

Rule 14: Lines longer than $\lambda/10$ (λ wavelength of clock frequency) that have not been matched should not be used.

Rule 15: Use small surface-mount devices (SMD) where possible, as they have a much better RF response. Keep line lengths associated with devices as short as possible to reduce series inductivity. Standard capacitors with interconnecting leads exhibit their first natural resonance at approximately 80 MHz.

Rule 16: Line drivers with symmetrical output (differential signaling; the signal difference is based on the 0 V potential) improve signal integrity, significantly cut interference emission and enhance immunity.

Rule 17: If several ICs are to be driven from one logic output, for example, routing a clock signal to a number of devices, split the trace near the devices (Fig. 6.44). The common line can be matched to its characteristic impedance at the distribution point, if needed.

Rule 18: Input and load capacitances should be as low as possible. This helps reduce charging currents for the state change, which in turn reduces magnetic-field radiation and ground return currents.

Rule 19: You need to prevent demodulation issues in analog circuits. Most EMC issues in analog semiconductor devices are caused by RF-signal demodulation. Analog circuits must be stable and continue working during a high-frequency disturbance event to prevent demodulation. This can only be achieved with an input filter or a feedback circuit.

Rule 20: Categorize lines on the printed circuit board and system wiring according to their signals in a similar manner as at the system level (e.g., clock, power, digital logic, or analog), and use a different wiring path for each category. Lay the different categories as far apart as possible, near the chassis ground, and successively from susceptible/non-disturbing to non-susceptible/highly disturbing.

Rule 21: A low-stray field arrangement can easily be achieved by using multi-wire flat ribbon cables. Strap them to the chassis ground. If too much cable is needed to run the cables to chassis ground, or if no chassis ground is available, running a metal foil underneath the ribbon cable will bring a significant improvement (imaging principle).

Rule 22: Eliminate all EMC passive conductors in an electronic system.

Rule 23: The power switch should, if possible, be integrated in the line filter. Use an LED on the low-voltage side to indicate power on.

Rule 24: The following shielding rules should be observed when designing the shielding enclosure:

– Low-frequency electric fields (1 MHz max. at system level) can be shielded by a thin-walled metallic enclosure or a plastic enclosure with metal coating.
– Low-frequency magnetic fields (1 MHz max. at system level) require thick-walled metallic enclosures; highly permeable materials are needed for power frequency magnetic fields.
– The leaks (holes, slots) determine the shielding effectiveness more and more with increasing frequency. The perforation with the largest extension d determines the scale of the reduction in shielding effect. The shielding effectiveness reduces to 0 dB (i.e., no shielding) for a perforation with an extension d (in m) $\geq$ 30 m / f (in MHz) ("$\lambda/10$ rule").
– Since the largest extension (in one direction) determines the shielding efficiency of shields with perforations, many small holes are better than a few large ones, and circular apertures are better than square ones.

Rule 25: The overall goal of all EMC measures for electronic systems should be to strengthen the electronics so that the emission limits and immunity requirements for the electromagnetic surroundings "budget" meet EMC standards without the need for an additional shielding enclosure.

6.6.3 Designing System Cabinets

After providing basic rules for EMC-compliant design of PCBs and shielding enclosures, let us now consider the "upper end" of possible system levels, a modular system cabinet, as depicted in Fig. 6.45.

As a material, aluminum is generally unsuitable for the enclosure, as it provides no shielding against low-frequency magnetic fields (see Sects. 6.4.2 and 6.4.3). Use round vents $d \leq 5$ mm in diameter instead of ventilation slots for effective shielding against high-frequency fields. Fit large apertures, for displays, for example, with metal screens or honeycomb windows (see Fig. 6.35 and Sect. 6.4.6).

Do not route the AC power supply through the cabinet, as it is difficult to filter out the low-frequency ripple voltages. Fit the line filter near the power cable entry point.

Connect all metal parts, such as cabinet and racks, with the protective conductor (PE, protective earth) as protection against contact voltages. No compensating currents should flow through these parts.

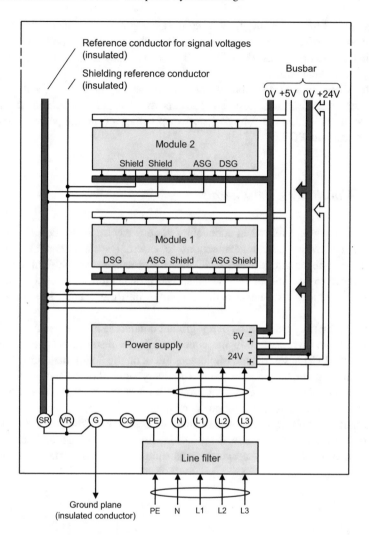

Fig. 6.45 Power supply and grounding in a system cabinet (*L1, L2, L3, N, PE* three-phase system, *CG* chassis earthing screw, *G* system ground, *SR* shielding reference conductor, *VR* voltage reference conductor, *Shield* shielding, *DSG* digital signal ground, *ASG* analog signal ground)

The system ground G is the system reference potential. Avoid using the system protective conductor PE as system ground, as it could be the same as the neutral conductor N and can be highly "contaminated". Instead, and if available as well as the system protective conductor, use a system ground that is clear of disturbance voltages. Insulate the cable to the ground plane well.

Reference conductors for signal voltages and shielding should be connected to the system ground. They should take the form of low inductance and low-resistant surfaces, a busbar or cables. Insulate them from metal parts in the enclosure and in the assemblies to prevent the flow of compensating currents that would neutralize the shielding effect of the metal parts. Run the reference conductors for the shielding or ground potentials individually and insulated to the reference point for voltages in critical cases.

Reference conductors should not be used as return conductors for the module power supplies. They should be shielded and twisted pairs or laminated busbars. You may need to filter out "stray" disturbance voltages from the power supplies in modules, such as PCB.

6.6.4 Connecting Peripherals

Great care must be taken when connecting various electronic systems, because the cables that provide the interconnections to external systems can also be very susceptible to disturbances. Here, EMC measures focus on the cable shields and the points where cables join to the systems. The propagation of wanted signals, the reduction of emissions, and also the couplings into the wanted signals must be considered.

When using plug connectors, every ground conductor in an interconnect (cable) should have its own contact in the connector. If the distance between cable and circuit on the PCB is significant, implement traces on the board that mirror the cable design: Lay a ground conductor parallel to every signal line on the printed circuit board, and connect with the system ground only at the circuit. Always twist the signal and associated return (ground) conductor in the cable.

If possible, fit all input and output lines with electrical isolators, like optocouplers or coupling transformers (Fig. 6.46). Conductors for ground and supply voltages and for shielding the two systems should not be connected together.

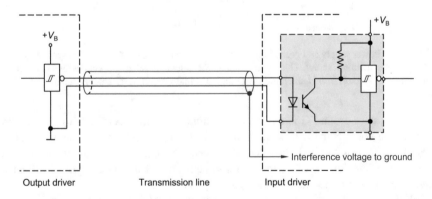

Fig. 6.46 Electrical isolation (galvanic separation) of two systems with an optocoupler

References

1. Directive 2014/30/EU of the European Parliament and of the Council of 26 February 2014 on the harmonisation of the laws of the Member States relating to electromagnetic compatibility, online: http://data.europa.eu/eli/dir/2014/30/oj
2. A. J. Schwab, W. Kürner, *Elektromagnetische Verträglichkeit*, Springer, 2010
3. H. W. Ott, *Electromagnetic Compatibility Engineering*, 1st edition, Wiley, 2009
4. K. H. Gonschorek, R. Vick, *Electromagnetic Compatibility for Device Design and System Integration*, Springer, 2009
5. C. R. Paul, *Introduction to Electromagnetic Compatibility*, 2nd edition, Wiley, 2006
6. *IEEE Std 100*, IEEE Standards Association, 1996
7. K. L. Kaiser, *Electromagnetic Shielding*, CRC Press, 2006
8. K. L. Kaiser, *Electrostatic Discharge*, CRC Press, 2005
9. Tecknit, *Electromagnetic Compatibility Guide*, 1998

Chapter 7
Recycling Requirements and Design for Environmental Compliance

Faced with a decreasing abundance of resources, together with ravaging environmental degradation caused by mounting waste and hazardous substances, we as engineers are forced to reconsider our approach to the design and development of electronics systems. The work of the development engineer no longer ends with the finished product, but must now also include the statutory responsibility for its *recycling*. This involves dealing with product recycling—the reuse and further use of the product, and/or material recycling—the reuse and further use of its constituent materials. *Design for recycling* must be our new guiding paradigm.

We begin this chapter (Sect. 7.1) with a discussion of the importance of a *circular economy*, in which products are designed to circulate in the production system without entering the environment. We next describe the circular economy's effect on the manufacture, usage, and disposal of electronic systems in Sect. 7.2.

Section 7.3 explores the concept of *product recycling* during the disposal process, including new marketing and design strategies for a more intensive product usage.

The materials in every electronic system must be disposed of at the end of their useful life. The commercial and ecological aspects of the necessary *material recycling* (Sect. 7.4) are determined by how well the system has been designed for disassembly and by the suitability of its constituent materials for recycling. The development engineer ensures the former with his/her system layout and the latter by the selection of materials. Material recyclability is thus established at the design stage by designing for disassembly and choosing suitable materials. We will explore the applicable principles and guidelines in Sects. 7.5 (disassembly) and 7.6 (suitable materials) in detail.

Section 7.7 concludes with a review of recommendations for electronic system development that is geared for environmental compliance.

© Springer International Publishing AG 2017
J. Lienig and H. Bruemmer, *Fundamentals of Electronic Systems Design*,
DOI 10.1007/978-3-319-55840-0_7

7.1   Introduction—Motivation and the Circular Economy

For hundreds of years now technical goods have been developed and produced solely with a view to their *use*, with little or no consideration as to what will happen to them after their useful life has expired. No thought was given to their possible reuse or recycling and to their final disposal. At the end of the product's life, it was simply thrown away—or at best, was brought to the dump. This was viewed as "disposal" in those days. Those were the days—and they still are to some extent in industry today—when the throwaway mentality held sway. The term "useless" was (and still is) used to refer to a to-be-disposed-of product's value, or lack thereof, in a technical or economical sense. Other possible residual product value (i.e., "usefulness"), such as materials and energy were, and still are, ignored. High product value is increasingly being lost after ever shorter product life spans and ends up as waste in a landfill. This throwaway mentality is now an existential threat to humankind because of the following:

- The exponential growth in the consumption of finite natural resources,
- The increasing release of hazardous substances into the environment with limited absorption capacity,
- The increased depletion of raw material resources by dumping products that have become obsolete in landfill sites (this makes recovery practically impossible as low volumes of materials are distributed in a dissipative manner).

The irrevocable loss of raw materials due to their increase in *entropy* is of particular concern. Entropy is a measure of the "disorder" of a system, that is, the number of different microscopic states a system can be in.[1] (The term "microscopic states" means the exact states of all the molecules making up the system.) Hence, the more "ordered" or "organized" a system is, the lower the microscopic disorder of this system, and the lower its entropy. This means that the entropy change during a process not only represents the difference in the quantity of substances existing at the beginning and end of the process, but also the change in the order of the involved substances from beginning to end.

Figure 7.1 shows that order increases initially with the extraction of valuable materials (substances) from the raw material base. In other words, we see a reduction in entropy because the materials' microscopic disorder is reduced. (Essentially, the properties of the extracted substances are more constrained then that of the raw material "mix".) The level of disorder then rises continually with the

---

[1]Complex building structures are good examples to illustrate the concept of entropy. Here, for example, building blocks that have been used to construct a wall are "highly organized" (i.e. they are arranged in a complex structure) and are thus in a *low-entropy* state. This state is achieved only by the input of energy. If this structure is left unattended, it will decay after a number of years, and the disorganized, high-entropy state will return (i.e., an unorganized heap of blocks). Generally speaking then, entropy is maintained, or it increases, in all natural processes.

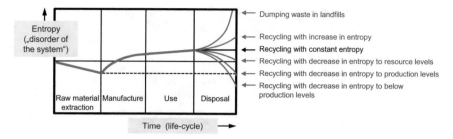

Fig. 7.1 Plot of the entropy of materials in the product life cycle where nowadays we convert low-entropy energy and materials into high-entropy waste. Product life cycles should take entropy retention into account to counter the irreversible loss of valuable raw materials due to the second law of thermodynamics

production of waste in manufacturing. The increase in disorder is further enhanced by "diluting" or "mixing" materials in the assembly of complex structures, followed by wear and tear and corrosion during the usage period. Finally, there is an exponential increase in disorder arising from the mixing of materials in landfills ("Dumping waste in landfills" in Fig. 7.1). The dissipative distribution of valuable substances renders them practically useless, they are effectively "lost forever." As a result of this directionality of the entropy law, the world's mineral resources are becoming scarcer and scarcer, increasingly limiting our economic prospects.

To counter this man-made "acceleration" of the second law of thermodynamics,[2] we must finally take into account the irreversibility of using inputs from the natural environment, thus, aiming for recycling with entropy retention as indicated in Fig. 7.1 ("Recycling with decrease in entropy").

Before proceeding, we first observe some facts and trends that will provide insight and motivation for improved recycling of electronic systems, with regard to their fabrication, usage, and disposal:

- The current widespread and large-scale use of electronic systems (e.g., there are currently about 4.5 billion cell phones in use worldwide),
- High rates of upgrading and disposal for "old" systems (e.g., the average life span of smartphones is currently about 4 years in the USA),
- The massive amount of electronic waste produced worldwide (the total amount of e-waste generated in 2018 is approximately 50 million metric tons),
- The large quantities of materials used in systems and the high concentration of materials during manufacture (one metric ton of materials must be sourced, transported, and processed to produce a "4 kg electronic system"),
- The high proportion of both hazardous materials and recyclables in the systems,
- The typically low reuse and recycling rates (30–40% in Germany),

---

[2]The second law of thermodynamics states that the universe evolves such that its total entropy always stays the same or increases.

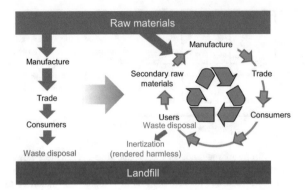

Fig. 7.2 Pivoting from a linear economy based on "take, make, dispose" material flow chain (*left*) that depletes finite raw materials and creates products that end up in landfills, to a regenerative circular economy (on the *right*). Here, products are designed to circulate with high quality in the production system, without entering the environment

– Significant energy consumption in standby mode (which represents 5–10% of all residential power used in most developed countries),
– Complex systems in particular are typically not designed for recycling.

Clearly, there needs to be a tectonic shift away from rampant throwaway consumerism to a resource-efficient and ultimately regenerative *circular economy*. We need to move from a linear "take, make, dispose" material flow (Fig. 7.2 left) to a circular economy framework by reusing or recycling products (Fig. 7.2 on the right). This will allow the deployed raw materials to be fully reintegrated in the manufacturing process beyond their so-called useful life.

The circular concept is grounded in the study of feedback-rich, nonlinear systems, particularly living systems. It considers that our products should work like organisms, processing "nutrients" that can be fed back into the cycle, whether biological or technical.

Lawmakers acknowledge the need to respond to these issues. *Closed Substance Cycle Waste Management Acts* have been enacted in most European countries to promote the circular economy model. The basis of this legal framework is the Waste Framework Directive of the European Union [1], which defines the primary waste-related terms, introduces a five-step waste hierarchy, and contains key provisions for national waste disposal laws. The purpose of this legislation is to promote the circular economy as a vehicle for conserving natural resources and to assure the environmentally compliant disposal of waste.

Section 3 of the German Closed Substance Cycle Waste Management Act entitled "Product Stewardship" stipulates that this duty of care applies to the entire life span of a product—from design and development through manufacture and deployment, and finally to disposal: "Products should be designed so that none, or a minimum of, waste is produced during their manufacture and operation, and that any waste generated at the end of their useful life is recycled or disposed of with minimum impact on the environment" [2].

Fig. 7.3 ARRE strategy for waste management, which prioritizes waste avoidance and aims to reduce unavoidable waste

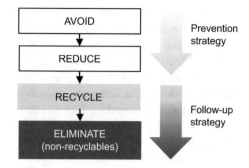

The responsibilities of the development engineer with respect to product stewardship are as follows:

1. To design, manufacture, and place products on the market that are durable and suitable for multiple reuse. It must be possible to properly and effectively reuse products in a safe and harmless manner and to disposed of them at the end of their useful lives with no negative impact on the environment.
2. To use, where possible, recyclable waste or secondary raw materials in the manufacture of products.
3. Labeling hazardous products so that any residual waste is recycled or disposed of in an environmentally safe manner at the end of their useful lives.
4. Affixing a label to the product w.r.t. return, reuse, and recovery options or responsibilities, and refund arrangements.
5. Taking back products and recovering or dispensing with any remaining waste after their recycling in an environmentally friendly manner.

Today, material recovery, in which the material is not altered, has priority over energy recovery from waste, that is, the extraction of the latent energy content. While, heretofore, non-recyclable waste was destroyed with as little impact on the environment as possible, the circular economy improves upon this with a preventative strategy for waste, rooted in *waste avoidance* and the *reduction* of *unavoidable waste*. A so-called ARRE strategy applies to waste management (Fig. 7.3).

7.2 Manufacture, Use, and Disposal of Electronic Systems in the Circular Economy

The greatest technical challenge posed by the circular economy is to develop environmentally compliant systems with the following characteristics:

– Waste is practically eliminated altogether during manufacture, usage, and disposal,
– Materials used in the products are almost completely recycled after usage.

The standards and criteria applicable to the circular economy also apply to the entire life cycle of an electronic system: to the manufacturing steps, to system operation and use, and to its disposal. The three relevant recycling loops for industrial waste, product, and material recycling are depicted in Fig. 7.4.

The manufacturing phase that is comprised of material processing and fabrication forms the *production waste recycling* loop (I). Industrial waste materials are fed back into, and reused in, the same production process. Because of the extensive use of chemical processes, every effort should be made during electronic assembly to deploy technologies that generate little or no waste. The aim here is to put in place closed manufacturing processes. As these issues are more in the realm of production and assembly than in the realm of electronic systems design, production waste recycling will not be dealt with further below.

There is typically no waste produced during the period of electronic system operation, so that there is no need for a recycling loop associated with system usage. That said, any ensuing waste material, such as data storage devices, should be integrated in higher level recycling loops. Energy losses during system operation and use, in the form of electromagnetic radiation and power consumption, for example, should be considered during design and development. Obviously, it is wise to design systems for minimum energy consumption. This includes the power consumed in standby mode. Here, for example, one energy-saving option is to have

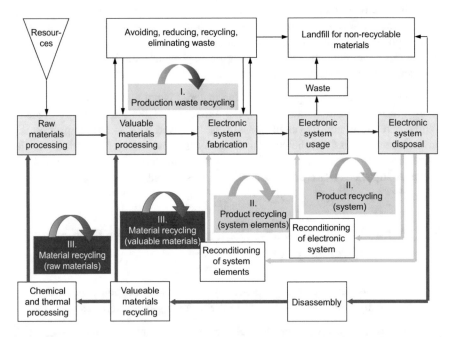

Fig. 7.4 Recycling loops for manufacture (production waste recycling) and disposal (product and material recycling) that electronic system design should aim for [3]

a power switch, which is easily accessible, and which allows the user to disconnect the system completely from the power source.

The system disposal process requires two recycling loops: *product recycling* (II) and *material recycling* (III).

The product (for product recycling at the system level) or product parts (for product recycling at the element level) are reused in the *product recycling* phase. This happens in one of two ways: (1) the product is used retaining the product functionality (reuse) or (2) the product continues to be used but with altered functionality (further use). The product design is unmodified, or only slightly modified, in both scenarios.

Product disassembly requires *material recycling*. Here, product materials are recovered and recycled in the material recycling loop. The materials are processed so they can flow back into the production process as valuable materials (valuable material recycling) for a similar (internal recycling) or different manufacturing process (further use). Alternatively, one can deconstruct them into raw materials in chemical and thermal processes and recycle them in a raw materials processing step (raw materials recycling).

Finally, we would like to point out that these recycling loops are supported by the maintenance operations covered in Chap. 4 (Fig. 7.5). While maintenance prolongs the life span of an electronic system (durability, see Sect. 7.3.2), recycling loops enable additional life cycles beyond the initial product. Both activities, while not always separable, are complementary and crucial in promoting ecological usage and disposal of electronic systems in a circular economy.

7.3 Product Recycling in the Disposal Process

If we take a closer look at the recycling loops in Figs. 7.4 and 7.5, we see that less value is extracted from goods as the loops increase in size, and the time and energy needed increase considerably, as do the costs (Fig. 7.6).

The smallest recycling loop, that is, product reuse, is therefore the best option (product recycling at system level, see Fig. 7.4). On the other hand, the least attractive approach is the recycling loop where the materials are completely dismantled and the raw materials are salvaged with an injection of energy (raw materials recycling, see Fig. 7.4). Despite this, material recycling is almost entirely the exclusive means of recycling employed for electronic systems at this time, where whole systems are disposed of after ever shorter life spans and the high value of commodities is lost.

This ecologically flawed trend dominates in the consumer goods space, where the useful lives (life spans) of products have been continuously eroded for years. This trend is less due to technology constraints than to artificially reduced life cycles driven by consumption-oriented marketing and advertising strategies (bereft of any moral basis). The problem is exacerbated by the manufacturer's policy of *planned obsolescence* or *built-in obsolescence*. Here, the useful life of electronics

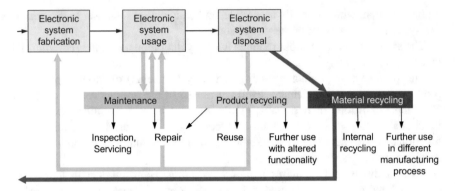

Fig. 7.5 Recycling loops in Fig. 7.4 are complemented by effective system maintenance (see Chap. 4) during the useful life of an electronic system

Fig. 7.6 Value changes and expenses for material and product recycling indicating that the latter (where the product is reused retaining the product functionality or with altered functionality) is not only ecological, but also economically the best option

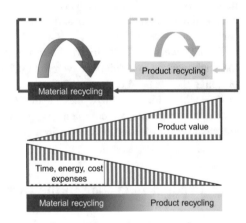

systems is deliberately cut, for example, by introducing "weak points" in the products such as using inferior materials in critical areas or suboptimal component layouts which cause excessive wear.

Materials requirements for technical products worldwide could be reduced to one-eighth (or 12.5%) of current levels (and thus by as much as 87.5%!) if the following complementary measures were taken:

– Doubling the average life span of all products,
– Doubling the amount of materials recovered,
– Halving the amount of materials used to manufacture every product.[3]

This purely theoretical estimation gives us some sense of the huge potential for the development of ecologically sensitive electronic systems! Recycling products

---

[3]This theoretical reduction to one-eighth (12.5%) is based on the threefold doubling of the savings ($2 \times 2 \times 2 = 8$).

along with effective maintenance programs to obtain longer useful lives for electronic systems will deliver the following benefits:

- A reduction in the quantity of materials processed per time unit,
- An increase in the material efficiency (number of operations per processed quantity of materials),
- A reduction in the amount of waste per time unit,
- Retaining value longer,
- Avoiding environmental impact due to transportation and packaging.

If one looks at the automobile sector where enormous benefits are accrued with second-hand cars and reconditioned spare parts, it is hard to understand why there is practically no product recycling in the electronics sector. However, there are a number of arguments against electronic product recycling:

- Technical issues: The longer usage periods work against technical developments that provide better functional, energy, and ecological value.
- Economical issues: Longer usage periods need reparability, a feature that is uneconomical in highly organized, cheap mass production, and cannot therefore be justified.
- Business reasons: The manufacturer is primarily interested in selling products and in short-term product liability periods; the manufacturer does not view itself as a "recycler", and instead chooses to leave this task to others.

Consequently, new marketing and design strategies are needed to drive product recycling in electronic systems. These are outlined in the following two Sects. 7.3.1 and 7.3.2.

7.3.1   New Marketing Strategy—Selling Usage

Electronic system users are principally interested in the benefits of their use. If usage is fully available, if maintenance and service are performed by third parties and the user is relieved of the responsibility for system disposal, he/she will not be particularly interested in purchasing the system, but will prefer instead to "*purchase the usage,*" more commonly known as *leasing* (Fig. 7.7). Leasing is worth looking at and promoting as an option for users of electronics systems, as it is already well established for office equipment, such as copiers.

7.3.2   New Design Strategy—Product Durability

The principle prerequisite for product recycling, i.e., product reuse, is its technical *and* moral durability which must also be accompanied by a willingness on the part of consumers to forego having the "very latest" version of a product when a prior

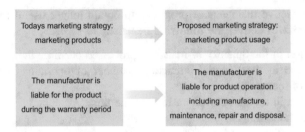

Fig. 7.7 New marketing strategy for electronic systems (products) that supports "purchasing the usage" instead of the product

version meets all of their needs. This is predicated by good *reparability* and *regenerability* in conjunction with maintenance (see Sect. 7.2). Every durable product must be capable of *adapting* to technical, technological, and design developments. Based on these three criteria, new design strategies for assuring durability, and thus product recycling, are as follows:

- *Design for durability,*
- *Design for regeneration,*
- *Design for adaptability.*

These design strategies are outlined in Tables 7.1, 7.2 and 7.3 in further detail.

Table 7.1 Strategy and principles of design for durability [3]

Design for durability	
Strategy	Dimensioning and designing systems for service over long periods with little or no maintenance
Principles and guidelines	– Avoiding wear and tear by electronic and optical means – Reducing wear and tear by compensating measures, adjustments and replacement of consumables – Preventing corrosion – Increasing reliability by functional and structural measures (functional elements with lower failure rates, introducing redundancy) – Deploying diagnostics with automatic error correction and scheduled maintenance (preventive maintenance)

Table 7.2 Strategy and principles of design for regeneration [3]

Design for regeneration	
Strategy	Dimensioning and designing systems so that simple and quick repairs can be carried out after a malfunction, and simple refurbishment or regeneration work can be carried out after loss or degradation of functionality, with the objective of a complete recovery to fully operational status
Principles and guidelines	Enabling ease of disassembly and reassembly of defective parts or parts for refurbishment (design for disassembly, see Sect. 7.5)

Table 7.3 Strategy and principles of design for adaptability [3]

Design for adaptability	
Strategy	Dimensioning and designing systems so that they are easy to upgrade to technical, technological and design changes with the objective of improving quality
Principles and guidelines	– Application of modular design principle (modular design approach or industry-wide standard module development, see Sect. 3.2.1) – Application of modular design principles by splitting the system into subsystems with long expected service lives and those that are expected to change or be extended, along with a clear separation of functional and design elements – Assuring the adaptability of modules by pre-emptive standardization – Long-term acceptance of the design solution by applying the design-for-usability principles of functionality, simplicity and authenticity

The third strategy, design for adaptability, is the most important. Durability and reparability are worthless if systems technologies or designs are outmoded. They have to be adapted to new technical, technological, and design developments, that is, they have to be able to be improved, upgraded, and updated. Design for adaptability is therefore the most challenging of the proposed design strategies: Engineers need to foresee and predict technical, technological, and design developments and trends. They also need to allow for future modifications to their designs so that systems can be easily adapted at a later stage.

Modularization is the key principle for design for adaptability (see Table 7.3). There is a greater variety of functions and structures in systems with adaptable modules that can be exchanged. Here are some examples of modular designs:

– Design of enclosures for electronics as 19-inch rack systems (see Sect. 3.2.1),
– Allotting spare slots for extensions in electronic systems, such as tuners, TV systems,
– Modular and scalable design of computer systems (PCs), which allow memory to be expanded, graphics cards to be upgraded, etc.

Product recycling at the element level (see Fig. 7.4) has the advantage over the system level in that components, and often modules, are largely standard items w.r.t. functionality and materials, and can thus be redeployed regardless of the product.

7.4 Material Recycling in the Disposal Process

Despite all its environmental and resource-saving benefits, product recycling only —but very effectively—postpones material recycling. The materials in every system (product) must be disposed of at some stage in the future, i.e., it will be deconstructed into its constituent materials and recycled in an environmentally compliant manner. As we discuss in this section, material recycling is thus the recycling procedure for breaking down a system into its material parts. It should have the following characteristics:

– Minimum disassembly effort,
– Maximum proportion of recyclables in good condition with minimum recycling effort,
– Minimum disposal cost for hazardous materials.

It is crucial that recyclables are not recouped with undefined characteristics, but in a standard quality that allows direct reuse as a valuable material (substance). The main elements of material recycling are as follows:

– Breaking down (disassembly),
– Salvaging recyclables,
– Disposing of non-recyclables, i.e., toxic materials.

Hence, the material recyclability of an electronic system is determined by its *suitability for disassembly* and the *suitability of its constituent materials*. Both must be assured in the design process by designing for disassembly and the suitability of constituent materials (Fig. 7.8). Of the two tasks, design for disassembly has priority, as systems that are easy to deconstruct into recyclables and waste materials facilitate and expedite the recovery and disposal of these two types of materials.

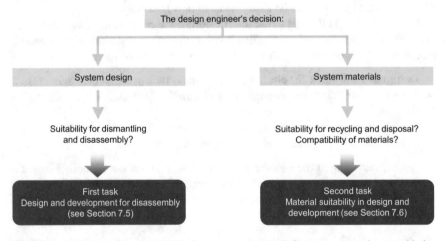

Fig. 7.8 Defining material recyclability for a system in the design process requires considering two different aspects, the suitability of the system's disassembly process (suitability for disassembly, see Fig. 7.9 top) and its constituent materials (suitability of materials, see Fig. 7.9 bottom)

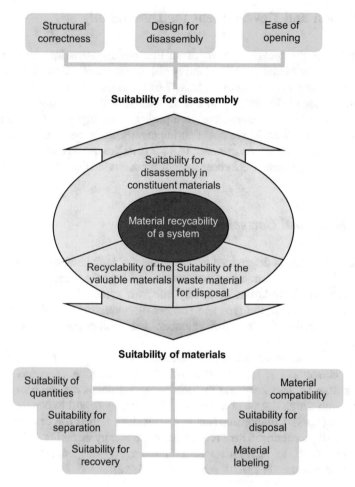

Fig. 7.9 Recyclability of a system w.r.t. material recycling is determined by its suitability for disassembly (Sect. 7.5) and the suitability of its materials (Sect. 7.6). Both can be split into nine more specific suitabilities that should be followed during design and development

We can divide the required suitability for disassembly and suitability of materials according to Fig. 7.9 into nine other sub-suitabilities. They define the standards and criteria that the development engineer needs to meet for designing and developing electronic systems for future recycling in an environmentally sensitive manner. Using these standards and criteria as our basis, we present the principles and guidelines for the development engineer in the following two Sects. 7.5 (design and development for disassembly) and 7.6 (material suitability in design and development).

7.5 Design and Development for Disassembly

Disassembly is the breaking down of an electronic system into its constituent elements. The combination of suitability for disassembly (discussed here) and suitability of materials (discussed in Sect. 7.6) determine the material recyclability of a system. Ease of disassembly should be assured by the three suitabilities below at the design and development stage in a project. Although maintenance is often considered a separate topic, we note that improvements in the ease of disassembly may also simplify system maintenance (e.g., if parts are more accessible and easily replaceable), which further enhances overall product viability.

7.5.1 Structural Correctness

A system's structure is determined by the number of individual structural elements including the type and number of connectors, and how the elements are arranged. The system design or layout decisively affects a system's suitability for disassembly. In this context, the type and number of connectors and element arrangements are critical—after all, their number and configuration can simplify or complicate disassembly.

When electronic systems are designed with disassembly in mind, a policy of structural correctness is required that includes the selection of *system assembly structures that enables easy and fast disassembly* without the need for specialist tools and with a high degree of separation (disassembly depth) in largely elemental components (parts). Components should be suited for further use or recycling with a minimum of processing. These requirements are met by assembly techniques that allow components to be disassembled in a largely independent sequential or simultaneous manner (Fig. 7.10, see also Sect. 3.2.2):

- Linear assembly techniques that support sequential disassembly,
- Hierarchical assembly techniques that support sequential and simultaneous disassembly,
- Radial assembly techniques that support simultaneous disassembly.

The *linear system assembly* structure consists of the standard *layered, stack,* and *sandwich assemblies*, where every part of the structure is put in place on a base or on the part put in place previously (see Fig. 7.10 top). This assembly type enables sequential disassembly.

Hierarchical system assembly consists of a structural base element B, on which the remaining structure is built (see Fig. 7.10 center). The base element can then be the reference point for disassembly, i.e., all dismantling operations will refer to this point in automated disassembly. Each structural element can be sequentially removed starting from the base element, and structural elements within each level in the hierarchy can be removed simultaneously, as well. A *nested* or *chassis system* is

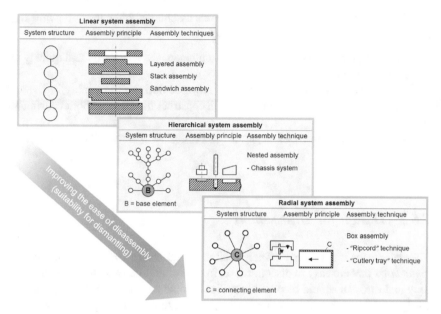

Fig. 7.10 Electronic system structures designed for disassembly should aim for assembling techniques that allow components to be disassembled sequentially or simultaneously

a good example of a hierarchical system assembly: Components that are largely independent of one another are placed in a level in the hierarchy in a load-supporting base element (frame, chassis).

The *radial system assembly* is the structure best suited for disassembly (see Fig. 7.10 bottom). This method allows you to disengage and separate all structural elements by disengaging or removing only one structural element, the connecting or locking element C. No special tools are needed and you can dismantle the system very quickly and without damaging any component parts.

The radial system method is mostly implemented as a *box assembly* (also known as a *shell* or *form-fit assembly*) where individual elements are fitted together and secured in the base (inserted) and held together by a central connecting (locking) element. The latter should be a requisite component anyway. Figure 7.10 (bottom) is a schematic of such a design: It pictures a round connector that can be taken apart by removing the main linkage C which is the outside casing. This conventional box assembly is often called the *"ripcord" technique.*

Another version of the box assembly is as a system comprising a lower part (base) and upper part (cover), both holding components in place. An example of this type of assembly is a cutlery tray, where the pieces of cutlery are held in place in recesses with a cover on top, hence the name *"cutlery tray" technique.*

The box assembly is suitable for electronic systems, as electronic components are only housed for reasons of convenience and not for operational reasons. (While positional tolerances can be tight, electronic functionality does not depend on the position of the electronic components.) Box assemblies offer the following recycling benefits:

- The electronics can be disassembled simultaneously very easily and quickly and at very low cost,
- No mechanical and thermal cutting is required during disassembly,
- Components and modules are easy to deconstruct for reuse without causing any damage,
- Easy to deconstruct and separate according to material types for ease of recycling.

7.5.2   Design for Disassembly

The design for disassembly principle aims for *assembly structures outfitted with connectors* that are easy to disconnect or, in case of non-detachable connections, are easy to destroy. It should be possible to disassemble the connector pairs quickly and easily, and there should be no need for specialist tools and equipment. Detachable form-fit-, form/press-, and press-fit connectors are recommended over detachable bonded connectors and connectors that can only be destroyed.

Snap-fit connectors (also: snap fits or snap-together connectors) as form-fit and form/press-fit connectors are easiest to dismantle (Fig. 7.11). This integral attachment feature can always be disengaged and should be designed so that it can be disconnected easily without hindrance (see Sect. 7.5.3).

Indirect snap-fit connectors fitted with separate clips are also suitable for disassembly. One or all of the clips can be destroyed during dismantling while maintaining the integrity of the connected parts.

Fig. 7.11 Split ferrite (for noise suppression in a cable) with a snap-fit connector. Snap fits are closed by pressing the interlocking component parts together

Fig. 7.12 Snap-fit
connectors designed for easy
opening with obvious
pressure point for opening
(*left*) and the appropriate
construction to ensure ease of
access (on the *right*)

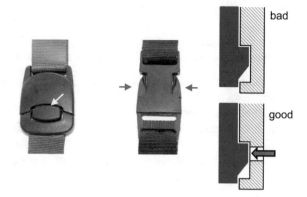

7.5.3   Ease of Opening

Form-fit-, form/press-fit-, press-fit- and bonded *connectors should be designed so
that they can be easily disconnected* without using specialist tools or technical
processes. The connection points should have the following characteristics:

–   Easily recognizable,
–   Easily accessible,
–   Protected against corrosion and soiling.

It is also crucial that the connection points are clearly visible. Let us take a
device enclosure consisting of an upper and lower part held together with a snap-fit.
If the snap-fit is not visible from the outside, it is not obvious where to press the
enclosure (from the outside) in order to release the snap-fit. This is because the
proper disengagement and separation of the enclosure parts has been obstructed.
This can be fixed by making the point at which pressure should be applied to open
the connector apparent, as well as the type of force that should be applied:
retractive, positive pressure or some other action (Fig. 7.12).

7.6   Material Suitability in Design and Development

In addition to the aforementioned suitability for disassembly (Sect. 7.5), suitability
of materials is a second critical aspect for the material recyclability of electronic
equipment. Choosing suitable materials forms the basis of economical and ecological material recycling by delivering a maximum proportion of recyclables at a
minimum recovery and disposal cost for constituent waste materials. This important
design goal is underpinned and sustained by the six suitabilities below.

7.6.1   Suitability of Quantities

Suitability of quantities is defined as the *minimization of the amount of materials* per system or component. This includes the following measures:

- Calculation/modeling software (e.g., FEM) for optimizing material fractions in structural entities,
- Lightweight design,
- Reinforcements and supports (frames, glass, corrugations, fins, sections, and the like),
- Inherent, fiber, and mineral reinforcement of plastics.

Everything we build doesn't have to be done extravagantly. This is especially true when it comes to protecting the environment and designing for recyclability: The less you use, the less has to be sourced, processed, fabricated, disassembled, salvaged, recycled, and disposed of.

Reinforcing plastics is a particularly effective measure for reducing the amount of plastic used; you can achieve the same, or better, characteristics with the right composite materials with lower plastic content than with unreinforced plastic. While the amount of reinforcing materials used is relatively low, it is the reduction in the amount of plastic used that is key.

Plastic parts are *inherently reinforced* by very uniform orientation of the molecular chains in the part. This is achieved in favorable flow conditions in the extruder and by controlling the pressure and temperature of the crystallization process. You can, for example, obtain a fourfold increase in tensile strength and an eightfold increase in Young's modulus with high-density polyethylene (HDPE) as against normal polyethylene (PE).

Glass-fiber-reinforced (GRP), carbon-fiber reinforced (CFK) plastics and plastics reinforced with Kevlar or aramid in fibrous, webbing and matting form are used for *fiber reinforcement*. The electrical and, especially, the thermal conductivity of carbon fibers are beneficial, as is the low coefficient of expansion of aramid. The latter is suited for thermally stressed parts.

Chalk, silicate, talc, mica, and calcium carbonate are composite materials used in *mineral reinforcement*. The mineral content is between 20 and 50%. Key benefits accrued are dimensional stability and heat resistance.

7.6.2   Suitability for Separation

The suitability for separation is determined by *how different materials are combined*, in particular how hazardous materials are combined with other non-hazardous materials in an electronic system or component, so that they can be quickly and economically broken down into material fractions with high purity levels. This design goal is achieved by the following:

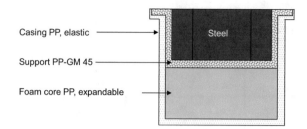

Fig. 7.13 Single-component system made of polypropylene (PP) of different strengths and textures on a base providing elastic support

- Using as few different types of materials as possible in a system or component,
- Avoiding the use of metal in plastic (insert technology) and plastic on metal (outsert technology),
- Avoiding the use of composite materials, especially surface coatings,
- Avoiding soiling,
- Using predetermined breaking points.

The ideal solution would be a one-component system that would obviate the need to separate the components for disposal. The worst type of solution is a composite system of a number of materials that cannot be disentangled. The materials cannot then be separated into pure material fractions. More practical solutions are often available between these two extremes. One, for example, could be a single-component system made of the same material with a varying texture so that it can perform different functions (Fig. 7.13).

While predetermined breaking points on packaging is standard today, e.g., heat-seal lids for yogurt cups and embossed tear strips for opening a package, their use for recycling electronic systems is still in its infancy.

7.6.3 Suitability for Recovery

Suitability for recovery guides the selection of *materials that can be recycled a number of times* at reasonable processing costs. Metals and thermoplastics are among such materials. Try to avoid using non-recyclable materials, like thermosetting plastics. (Thermoplastics can be remelted any number of times, whereas thermosetting plastics cannot be deformed again once they have been hardened.)

We don't just want to salvage and reuse materials from the consumption chain to halt the depletion of finite natural resources, we also want to avoid losing the valuable energy contained in the materials: For example, energy savings from recycling compared to the primary production of plastics is 94%; it is 68% for aluminum; 64% for carbon steel, and 43% for glass [4]. And we have, for example, 85% less CO_2 emissions from recycling aluminum compared to primary production [4].

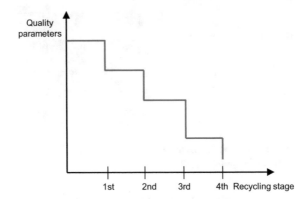

Fig. 7.14 Cascade principle with downcycling where materials go through a cascade of consecutive uses in which each successive use has less desirable properties. While this is considered only a short-term solution, it is nevertheless eco-efficient because it will save the materials and resources required to manufacture the downcycled products

Plastic recycling must be treated differently. The cascade principle depicted in Fig. 7.14 shows that the quality parameters worsen incrementally and significantly when mixed non-metallic materials are recycled. Paper can only be repeatedly recycled up to a maximum of seven times and plastic polypropylene (PP) up to five times while keeping the same quality parameters.[4] Afterward, the quality of these materials is degraded to such an extent that they can only be used for inferior products such as cardboard or flower tubs, etc. This type of recycling is called *downcycling*.

Upcycling is an effective countermeasure where waste materials, useless, or low-quality products are transformed into new materials or products of better quality. For example, when considering plastic products, material properties are almost fully maintained by adding the appropriate quantities of plastic recyclate. This upgrading of plastics by adding aggregates is also called *compounding*.

Coextrusion, i.e., the extrusion of multiple layers of material simultaneously to yield properties distinct from those of a single material, is an extremely economical technique for recycling mixed waste plastics. In this process, plastic waste is injection molded with virgin plastics. In Fig. 7.15, the inner part of the pipe wall is made from plastic waste; the inside and outside layers are injection molded with virgin plastic. The pipe looks "like new" in quality and finish, and its inside surface meets hygienic quality standards for new products, even in the food industry and in the field of medicine.

---

[4]After recycling office paper up to seven times, for example, the required long fibers are not sustainable anymore and only a "lesser use" with short fibers is possible, such as cardboard or toilet paper.

Fig. 7.15 Pipe constructed
by coextrusion where two or
more materials are pressed
through the same extrusion
head (die) to produce a single
part

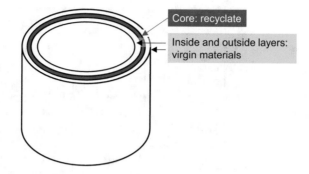

Core: recyclate

Inside and outside layers:
virgin materials

7.6.4   Material Compatibility

Compatibility issues arise when inseparable material combinations are unavoidable
in a system. In this case, only *combinations that are mutually compatible* and that
can be recycled economically and with high quality are approved. This is best
achieved with materials that have the same chemical origins and/or constituents.

Compatibility matrices are available for assessing the compatibility of different
metallic substances and plastics (Fig. 7.16). In the case of plastics, you should be
particularly careful with process temperatures, i.e., melting temperatures and temperatures at which damage occurs.

Compatibility issues caused by contamination of metals and plastics are of
particular concern. Such contaminations are as follows:

– Substances absorbed by the material that are very difficult to remove,
– Substances that come in contact with the material when in use and that cause
  irreparable damage to it (e.g., mineral oils or solvent for plastics),
– Finishing, galvanization, and electro-plating materials, especially plastics.

Finishing is a major contamination issue. Galvanization or electro-plating of
plastics are also critical and should be avoided. Thus, for example, a metallic cap
should be used, instead of a metalized plastic cap, as an electromagnetic shield for a
module.

7.6.5   Suitability for Disposal

Suitability for disposal optimizes the materials deployed in electronic systems in
terms of their *subsequent recyclability*. The main goals are as follows:

– Avoiding toxic substances that cannot be disposed of or whose disposal requires
  significant investment,

Matrix material \ Additive	PE	PVC	PS	PP	POM	SAN	ABS	PBTP	PETP	PMMA
PE	●	○	○	●	○	○	○	○	○	○
PVC	○	●	○	○	○	●	◉	○	○	●
PS	○	○	●	○	○	○	○	○	○	○
PP	⊙	○	○	●	○	○	○	○	○	○
POM	○	○	○	○	●	○	○	⊙	○	○
SAN	○	●	○	○	○	●	●	○	○	●
ABS	○	◉	○	○	⊙	○	●	◉	⊙	●
PBTP	○	○	○	○	○	○	⊙	●	○	○
PETP	○	○	⊙	○	○	○	⊙	○	●	○
PMMA	○	●	⊙	○	⊙	●	●	○	○	●

● compatible
◉ partially compatible
⊙ compatible in small amounts
○ incompatible

Fig. 7.16 Excerpt from a compatibility matrix for plastics

- If the use of toxic substances is unavoidable, they should be separated or used in closed units if possible,
- Preferential use of degradable materials.

The contamination and damage caused by harmful substances often only comes to light when they come in contact with the environment in an inappropriate manner. Hazardous substances when used as additives in a plastic, such as gallium arsenide (GaAs) in an electronic circuit or cadmium in a nickel-cadmium battery, typically have no effect on their surroundings and are thus harmless when in use.

Hazardous substances should be only considered with statutory approval in compliance with national regulations on hazardous materials, such as [5] in the United States or [6] in the European Union. Unavoidable hazardous substances must be used separately in electronic systems to facilitate their disposal at the end of their life spans. The removal of toxic electronic components from printed circuit boards before the boards are disposed of provides a good example.

Suitability for disposal dovetails well with the increased deployment of degradable materials that can be completely disposed of without impacting the environment. Among the methods for degrading materials are as follows:

- Biological decomposition by micro-organisms,
- Photodegradation by light,
- Chemical decomposition by oxidation or hydrolysis.

While degradable materials are used in some products, such as garbage bags, their use in electronic systems is still awaiting the green light. The typical long-term instability of these materials under operating conditions, especially when exposed to heat, light, pressure, and humidity, remains a significant barrier to their introduction in this space.

7.6.6   Material Labeling

Material labeling is a key requirement for disposing an electronic system in a way that is compatible with its materials. Electronic components and modules should be *marked to indicate their constituent materials and applicable disassembly and cutting mechanisms*. With this information, you can properly select, sort, fraction, and process the materials. Materials should have distinct permanent markings that can be scanned electronically. These markings should be subjected to norms, as well.

International *recycling codes* that assist with the detection of the physical composition of a product or system exist (Fig. 7.17). They consist of a recycling symbol and a number that characterize the material [7, 8]. Plastics are marked with the numbers 01–07, batteries with 08–14, glass with 70–79, and composites with the numbers 80–99. An abbreviation for the material group is usually specified, as well.

In order to identify more complex material systems, memory chips that cannot be removed and that can be read without any physical contact are available. Recycling information and data on the functional stress on the system while in service can be stored in these chips (life cycle assessment). Furthermore, you can accurately identify plastic materials by integrating very small combinations of fluorescent dyes in plastics for analysis with a fluorescence spectrometer. Near infrared spectroscopy (NIR) can also be used to identify substances.

7.7   Recommendations for Environmentally Compliant Systems

Electronic system design for recycling is based on a holistic assessment of manufacture, usage, and disposal. Recycling loops are facilitated at all stages of a system's life cycle. These loops consist of avoiding waste during manufacture; system design for maintenance and refurbishment while the system is in service;

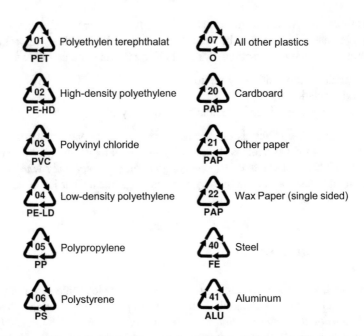

Fig. 7.17 Common recycling codes (excerpt) for different materials to facilitate easier recycling or other reprocessing. Materials are to be identified by a numeric code (mandatory) and/or abbreviation (voluntary). Some countries have adopted different codes which mainly differ in their granularity

and system design and development with a disassembly process in mind that is tailored for disposing the materials in the system.

Design process

The steps in the design and development process of an electronic system, where the development engineer makes decisions that impact the industrial waste, the energy consumption, the service life, the system layout, and the materials used in the system, are key. While service life is critical for product recycling, the suitability for disassembly and the suitability of constituent materials based on the layout and the deployed materials are critical for the material recyclability of the system at the end of its life cycle.

Choosing materials

When choosing materials, it is wise to restrict the number of different materials used, to ensure compatibility between the different materials, and to consider their ease of recycling when making selections. We recommend the use of thermoplastics (because they can be repeatedly melted down) over thermosetting plastics and self-reinforced plastics over those with reinforcing additives. You should try to avoid composites and surface coatings because they are difficult to break down into their component parts and thus inhibit recycling. Materials should be well marked and labeled so they can be easily sorted and segregated later at the end of the product's life span.

The development engineer should always try to minimize the use of materials when designing an electronic system. You can reduce the thickness of the sides of objects without loss of rigidity, for example, by using fins and honeycomb structures. Another recommended strategy is to reinforce plastics and rely on the inherent reinforcement of plastic parts to reduce the amount of plastic used. Recyclability greatly depends on how different materials are combined and joined: You should give priority here to press-fit and form-fit connectors, such as snap-fit connectors and those with adhesive bonds. If possible, do not use metal inserts in plastic and plastic on metal (outsert), as they are difficult to separate during the recycling process.

Disassembly and disposal

As disassembly costs are a high proportion of the total costs involved in material recycling, electronic systems should be configured for easy disassembly. An appropriate physical layout is essential here, which will ideally allow simultaneous disassembly, or at least sequential disassembly. Connection points should be easily identifiable and easy to access and dismantle, and the respective parts should not suffer damage when separated. Automated disassembly is also worth aiming for, with the provision of suitable machine tools and accessible surfaces between components and tools. You should plan how components are to be used after disassembly and incorporate this in the design. You should also allow for simple cleaning and refurbishment of these components.

The disposal of materials should be part of the requirements list for electronic system design. Use degradable materials where possible. If hazardous materials must be used, use them segregated in closed units so they can be separated for disposal. Avoid the use of hazardous materials that cannot be easily disposed of, or can be disposed of only at a high cost.

References

1. *Directive 2008/98/EC of the European Parliament and of the Council* of 19 November 2008 on waste and repealing certain Directives. Online: http://data.europa.eu/eli/dir/2008/98/oj
2. Closed Substance Cycle Waste Management Act (Kreislaufwirtschafts- und Abfallgesetz, KrW-/AbfG) of the Federal Republic of Germany, 27. September 1994, current version (2017)
3. G. Röhrs „Recyclinggerechte Fertigung und Gestaltung", chapter 3 in: W. Krause, *Fertigung in der Feinwerk- und Mikrotechnik*, Carl Hanser Verlag, 1996
4. *Recycling for Climate Protection*, Report of the Fraunhofer Institute for Environmental, Safety, and Energy Technology UMSICHT, February 8, 2011. Online at: http://www.alba.info/fileadmin/alba/pressemappe/recycling_fuer_den_klimaschutz/110210_CO2_Studie_ALBA_Group_final_v4.pdf
5. *Federal Hazardous Substances Act (FHSA)* of the United States Consumer Product Safety Commission (CPSC), current version. Online at: https://www.cpsc.gov/s3fs-public/pdfs/blk_pdf_fhsa.pdf

6. *Directive 2002/95/EC of the European Parliament and of the Council* on the restriction of the use of certain hazardous substances in electrical and electronic equipment, current version. Online at: http://data.europa.eu/eli/dir_del/2015/863/oj
7. *Christie Engineering Standard – Packaging Labeling and Design for Environment Guidelines,* Includes lists of material codes in several countries. Online at: https://www.christiedigital.com/Documents/Supplier%20Documentation/Packaging-Spec-010-101136-02.pdf
8. *European Parliament and Council Directive 94/62/EC* of 20 December 1994 on packaging and packaging waste, current version. Online at: http://data.europa.eu/eli/dir/1994/62/oj

Chapter 8
Appendix

This appendix is intended to introduce the reader to the types and formats of technical documentation that are typically encountered in electronic system designs. This encompasses geometric dimensioning and tolerancing in technical drawings (Sect. 8.2), preferred numbers (Sect. 8.3), and schematic symbols for electronic components (Sect. 8.4), including their labeling with colors and characters (Sect. 8.5).

8.1 Notes and Rules on Technical Drawings

A complete set of drawings always contains a general drawing for an electronic system or module and a bill of components. Individual assembly drawings with more detailed views may be required. In addition, detail part drawings are needed as well for all components to be produced.

8.1.1 Title Block

Every detail part drawing should have a title block with the information as stipulated by ISO 7200 (Technical product documentation—Data fields in title blocks and document headers). It is placed at the bottom, usually the right-hand corner of the drawing. Two examples are shown below:

© Springer International Publishing AG 2017
J. Lienig and H. Bruemmer, *Fundamentals of Electronic Systems Design*,
DOI 10.1007/978-3-319-55840-0_8

Responsible dep.	Technical reference	Creator		Approval person		
		Jason Miller				
TU Dresden		Document type		Document status		
		Single part drawing		First issue		
Institute of		Title, additional title		Identification no.		
Electromechanical and				Ifte2018-01.001		
Electronic Design		Spacer		Rev.	Date of issue	Sheet
					19 July 2018	1/6

General tolerances						
ISO 2768-mK						
Workpiece edges		Material S185		Scale	1:1	
ISO 13715		Semi-finished material		Mass: approx.	1kg	
Dept.	Technical reference	Creator		Approval person		
		Jason Miller				
TU Dresden		Document type		Document status		
		Single part drawing		First issue		
Institute of		Title, additional title		Identification no.		
Electromechanical and				Ifte2018-01.001		
Electronic Design		Spacer		Rev.	Date of issue	Lang. Sheet
					19 July 2018	en 1/6

8.1.2   Scales

Scales to increase or decrease the size of a part in a drawing are defined in ISO 5455 (Technical drawings—Scales), where the 1:1 scale represents the actual size of the part.

Reduction			Real size	Enlargement		
1:10	1:5	1:2	1:1	2:1	5:1	10:1
1:100	1:50	1:20		20:1	50:1	100:1
etc.	etc.	etc.		etc.	etc.	etc.

8.1.3   Identification Number

The identification number of the drawing should be encoded. A typical format for the number might be comprised of the product name, module number, and a part item number. Example: Ifte2018-01.001.

Ifte2018-	01.	001
Product name	General drawing: 00 Module drawing: 01–99	First single-part drawing for module 01

8.1.4 Paper Sizes

Paper sizes for technical drawings correspond to the formats in the primary series A according to ISO 216 (Writing paper and certain classes of printed matter—Trimmed sizes—A and B series, and indication of machine direction).[1]

Paper in the A series format has a $\sqrt{2} \approx 1.414$ aspect ratio,[2] rounded to the nearest millimeter. A0 is defined so that it has an area of 1 m^2 before rounding. Successive paper sizes in the series (A1, A2, A3, etc.) are defined by halving the length of the preceding paper size, so that the long side of the next smaller format, i.e., A$(n + 1)$, is the same length as the short side of the current format, i.e., An, before rounding.

The most frequently used size in this series is A4, which is 210 mm × 297 mm (8.27 in × 11.7 in). For comparison, the letter paper size commonly used in the USA and Canada (8.5 in × 11 in or 216 mm × 279 mm) is approximately 6 mm (0.24 in) wider and 18 mm (0.71 in) shorter than A4.

Format A4 is approved only as a portrait format in standards and the other sizes A3–A0 only as landscape formats (ISO 5457, Technical product documentation—Sizes and layout of drawing sheets).

Paper sizes larger than A4 should be folded so that the title blocks may be stacked on top of one another and are easily readable.

8.1.5 Line Styles and Widths

When selecting a line width, readability should be the main criterion; however, wide lines should not be used excessively in complex drawings. The most commonly used line styles and line widths as well as their respective applications are listed below.

---

[1]The USA and Canada do not use the ISO paper sizes; instead they use the Letter, Legal and Executive sizes. Although they have also officially adopted the ISO 216 paper format, Mexico, Panama, Venezuela, Colombia, the Philippines, and Chile also primarily use U.S. paper sizes.

[2]The geometric rationale behind the square root of 2 is to maintain the aspect ratio of each subsequent rectangle after cutting or folding an A series sheet in half, perpendicular to the larger side.

Line styles	Line widths in mm		Applications
_____ Continuous line (wide)	0.50	0.70	– Visible edges and contours – Screw tips – Used thread lengths
_____ Continuous line (thin)	0.25	0.35	– Dimension lines, extension lines – Hatching – Bending lines – Leader lines
Irregular continuous lines and zigzag lines	0.25	0.35	– Boundaries for fragmentary, broken and sectional views
– – – – – – – – Broken line (thin)	0.25	0.35	– Hidden object edges and contours
– · – · – · – · – · Dot-dashed line (thin)	0.25	0.35	– Center lines – Symmetry axes – Pitch circles for gear teeth – Bolt hole circles
▬ · ▬ · ▬ · ▬ Dot-dashed line (wide)	0.50	0.70	– Showing cross-sectional planes

8.1.6   Sectional Views

Sectional views show an object's hidden features. The following basic rules apply to sectional views (ISO 128-40, 44, 50, Technical drawings—General principles of presentation—Part 40: Basic conventions for cuts and sections, Part 44: Sections on mechanical engineering drawings, Part 50: Basic conventions for representing areas on cuts and sections):

– Cut surfaces are hatched.
– Cavities are not hatched.
– All cut surfaces in a part are hatched in the same manner and direction.
– At least one non-sectional view is generally required with sectional views (rotating parts are an exception).
– Intersecting lines are drawn in non-sectional views, if the section is not clearly visible; arrows indicate the direction of projection.
– Different types of materials are hatched differently (ISO 128-50).
– Hatching directions and angles vary for multiple parts with the same material.
– Complete workpieces, such as pins, screws, and key ways, are drawn uncut in longitudinal section.

Commonly used sectional views are shown in the three figures below. The *full section view* cuts through the whole part. A *half-section* is suitable only for symmetrical elements. *Broken views* (interrupted views) are used to shorten a longer part. The part between the break lines should not contain any other details, such as holes and steps. *Sections* (cutting paths or cutting-plane lines) can be used to increase the concentration of details in a sectional view. They should be marked with letters at the vertices if the cutting path is not readily apparent.

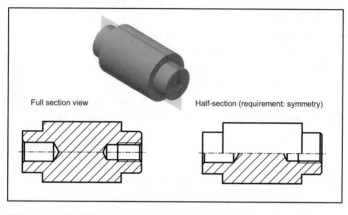

Full section view   Half-section (requirement: symmetry)

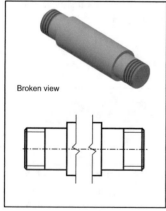

Broken view

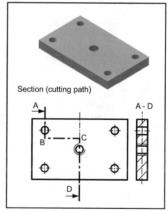

Section (cutting path)

Other sectional views:

– Partial section view: shows only part of an object.
– Profile section: sectional view restricted to cut surface, any hidden contours are ignored.

Details can be scaled for emphasis.

8.2   Geometric Dimensioning and Tolerancing

Dimensions define the nominal geometry and allowable variations, i.e., tolerances. Dimensions are normally specified in mm (unit not written), and series $R'20$ standard measurements (see Sect. 8.3) are preferred. They are used in views where it is clear what they refer to. External dimensions are always specified. Other rules to be followed are:

– Dimensions should only be written once per part.
– If possible, hidden object edges are not dimensioned.
– Dimensions are specified outside the single-part view; they are written within the single-part view in exceptional cases only.
– Intersecting dimension lines or extension lines should be avoided; chain dimensions and chain dimensioning should be avoided, too.

8.2.1   Elements of Specified Dimensions

Specified dimensions contain extension lines, dimension lines, dimension line terminations, and measurement values with tolerances.

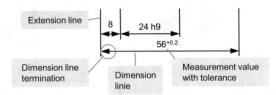

Element	Description
Extension line	Extension of geometry, narrow continuous line, located 2 mm beyond dimension line
Dimension line	Continuous narrow line, 10 mm from object, 7 mm distance between dimension lines
Dimension line termination	Preferably a full-headed arrow
Measurement value with tolerance	Dimension and tolerance values in mm (unit not written); the reading direction is from the bottom up or as seen from the right (w.r.t. the title block); the measurement value is written above the dimension line

8.2.2   Dimension Types

Production-based dimensioning is the primary method used. *Function-based* dimensions are used only if a module function is thereby assured. The purpose of *test-based* dimensions is to verify a given geometry, such as distances between holes, and hence they typically appear in custom inspection drawings.

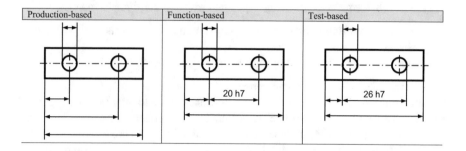

8.2.3   Tolerance Terminology

All dimensions must have a tolerance. Every feature on every manufactured part is subject to variation, therefore, the limits of allowable variation, i.e., tolerances, must be specified.

A dimension of the produced part may not be bigger than a *maximum dimension* nor smaller than a *minimum dimension*. The difference between the maximum dimension and the nominal dimension is the *upper deviation*, and the difference between the minimum dimension and the nominal dimension is the *lower deviation*. The difference between the maximum and minimum dimensions or between the upper and lower *deviations* is called the *tolerance*.

Schematic	Symbol	Term	Example: $8^{+0.25}_{+0.05}$ (mm)
	N	Nominal dimension	8.00
	G	Maximum dimension	8.25
	K	Minimum dimension	8.05
	C	Mean tolerance value	8.15
	E_C	Mid tolerance value	0.15
	ES, es	Upper deviation	0.25
	EI, ei	Lower deviation	0.05
	T	Tolerance	0.20

Note ES, EI are for internal dimensions (e.g., holes) and es, ei for outer dimensions (e.g., shafts)

8.2.4 *Engineering Tolerances*

Detail part drawings contain the following types of tolerances:

– *Dimensional tolerances,* comprising the following:

  • General tolerances (specified in the title block),
  • ISO tolerances (specified with nominal dimensions, e.g., 24 h9),
  • Custom toleranced dimensions (specified with nominal dimensions, e.g., $56^{+0.2}$),

– *Form and positional tolerances* for deviations from forms and installation positions,
– *Surface specifications* for workpiece surface conditions.

General tolerances are recommended, ISO tolerances are only for functional or part-fit requirements. Custom toleranced dimensions should only be used if general and ISO tolerances are unsuitable, for example, for a coarse prototype.

A *datum* is a virtual ideal plane, line, point, or axis. These are referred to by one or more "datum references" which indicate measurements that should be made with respect to the corresponding datum feature.

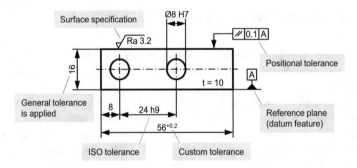

8.2.5 *General Tolerances*

The title block contains the general tolerance details applicable for all dimensions that are not explicitly toleranced. The referenced ISO standard with the tolerance classes to be applied is binding. The tolerance classes define the size and position of the tolerance zones as a function of the dimension size.

ISO 2768	Categories	Applications	Categories	Applications
Fine	f	Lengths, curves, bevels, angles	H	Straightness, surface flatness, perpendicularity, symmetry, true run, axial run
Mean	m		K	
Coarse	c		L	
Very coarse	v			

8.2.6 ISO Tolerances

ISO tolerances are composed of one letter and one number that follow the nominal dimension. They specify the position (letter) and size (number) of the tolerance zone as defined in ISO 286 (Geometrical product specifications—ISO code system for tolerances on linear sizes).

Example: 6 f7		Explanation
Base line (6 mm)	$f \triangleq -10$ µm from the base line for a nominal dimension of 6 mm	Position of the tolerance zone: lower case letters for outer dimensions, capital letters for internal dimensions
	Quality $7 \triangleq 12$ µm tolerance range for a nominal dimension of 6 mm	Extension of the tolerance zone

8.2.7 Form and Positional Tolerances

Form tolerances specify the approved range for the object to deviate from the ideal form. Positional tolerances define the allowed deviations of the locations of at least two elements w.r.t. one another.

Form tolerances		Positional tolerances	
Examples	Symbols	Examples	Symbols
Surface flatness	⌗ 0.1	Parallelism (between surface and reference surface)	A // 0.1 A
Straightness	— 0.1	Perpendicularity (between line and reference surface)	A ⊥ 0.1 A
Circularity	○ 0.1	Axial run (between surface and reference axis)	A ↗ 0.1 A
Cylindricity	⌭ 0.1	True run (between line and reference line)	A ↗ 0.1 A

8.2.8 Surface Specifications

Specifications for custom surfaces are given in the detail part drawing and are specified above the title block for the remaining surfaces. The roughness should, if possible, be specified by the averaged surface roughness Rz (mean value of the surface roughnesses from five neighboring test distances) or the mean roughness Ra (arithmetic mean value of the absolute values of the roughness profile from the mean line within the test distance); both values are specified in μm.

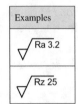

Symbols	Manufacturing process	Examples of realizable production qualities	Ra in μm
✓	unrestricted	Turning	≥ 0.4
		Drilling	≥ 3.2
ⱱ	material removed	Milling	≥ 1.6
		Longitudinal grinding	≥ 0.025
ⱷ	no material removed		

Examples:

√ Ra 3.2

√ Rz 25

8.2.9   Material Specifications

The material of a part is specified in the title block and, if necessary, in the bill of components, either as a material code (e.g., S185) or a material number (e.g., 1.0035).

8.3   Preferred Numbers—Renard and E-Series

Preferred numbers are standard guidelines for choosing exact product dimensions and other numerical characteristics. They support useful graduations of, for example, screw diameters, resistance values, and motor ratings, and thus, increase the likelihood of compatibility between different systems through the use of, e.g., parts and connectors that are a "standard size". Since the dimensions are equally spaced on a logarithmic scale, they also help to minimize the number of different sizes that need to be manufactured or kept in stock. Preferred numbers are defined in ISO 3 (Preferred numbers—Series of preferred numbers).

Preferred numbers are rounded elements in geometric series, i.e., the factor between two consecutive numbers is approximately constant. The numerical values are derived from the logarithmical splitting of decades; hence, the numbers are equally spaced on a logarithmic scale. The number of elements in each decade is constant, i.e., there are the same number of preferred numbers between 1 and <10 as there are between 10 and <10^2. This number of elements within a decade is called the *step number r*. The aforementioned constant ratio between two sequential preferred numbers is defined as *graduation* q_r (derived from the step number r with $q_r = \sqrt[r]{10}$).

The so-called *Renard series*[3] R5, R10, R20, R40, where the step number $r = 5$, 10, 20, 40, are *basic series*. The graduations in a Renard series are approximately the 5th (R5), 10th (R10), 20th (R20), or 40th (R40) root of 10 (approximately 1.58, 1.26, 1.12, and 1.06, respectively).

The Renard series can be rounded to produce the *rounded series R′ and R″*, where R″ is the most rounded series (it should not be used if possible). The values in the rounded series R′ are used as *standard sizes*, e.g., as preferred values for measures of length.

Basic/Renard series Main values				Rounded series Rounded values			Basic/Renard series Main values				Rounded series Rounded values		
R5	R10	R20	R40	R′10	R′20	R′40	R5	R10	R20	R40	R′10	R′20	R′40
1.0	1.0	1.0	1.0	1.0	1.0	1.0		3.15	3.15	3.15	3.2	3.2	3.2
			1.06			1.05				3.35			3.4
		1.12	1.12		1.1	1.1			3.55	3.55		3.6	3.6
			1.18			1.2				3.75			3.8
	1.25	1.25	1.25	1.25	1.25	1.25	4.0	4.0	4.0	4.0	4.0	4.0	4.0
			1.32			1.3				4.25			4.2
		1.4	1.4		1.4	1.4			4.5	4.5		4.5	4.5
			1.5			1.5				4.75			4.8
1.6	1.6	1.6	1.6	1.6	1.6	1.6		5.0	5.0	5.0	5.0	5.0	5.0
			1.7			1.7				5.3			5.3
		1.8	1.8		1.8	1.8			5.6	5.6		5.6	5.6
			1.9			1.9				6.0			6.0
	2.0	2.0	2.0	2.0	2.0	2.0	6.3	6.3	6.3	6.3	6.3	6.3	6.3
			2.12			2.1				6.7			6.7
		2.24	2.24		2.2	2.2			7.1	7.1		7.1	7.1
			2.36			2.4				7.5			7.5
2.5	2.5	2.5	2.5	2.5	2.5	2.5		8.0	8.0	8.0	8.0	8.0	8.0
			2.65			2.6				8.5			8.5
		2.8	2.8		2.8	2.8			9.0	9.0	9.0	9.0	9.0
			3.0			3.0				9.5			9.5
							10.0	10.0	10.0	10.0	10.0	10.0	10.0

Each of the Renard sequences can be reduced to a subset by taking every *n*-th value in a series, which is designated by adding the number *n* after a slash, for example, R20/3 (aperture values of a camera).

The series can be extended upwards or downwards by multiplying by …0.01; 0.1; (1;) 10; 100 …. Due to this repetition after every 10-fold change of the scale,

---

[3]The French army engineer Charles Renard proposed in the 1870s a set of preferred numbers in order to significantly reduce the number of different sizes of balloon ropes the French army kept on inventory. He divided the interval from 1 to 10 into 5, 10, 20, or 40 steps, based on a constant factor between two consecutive numbers (the 5th, 10th, 20th, or 40th root of 10).

they are particularly well suited for use with SI units. It makes no difference whether preferred numbers are used with meters or millimeters.

E6	E12	E24
1.0	1.0	1.0
		1.1
	1.2	1.2
		1.3
1.5	1.5	1.5
		1.6
	1.8	1.8
		2.0
2.2	2.2	2.2
		2.4
	2.7	2.7
		3.0
3.3	3.3	3.3
		3.6
	3.9	3.9
		4.3
4.7	4.7	4.7
		5.1
	5.6	5.6
		6.2
6.8	6.8	6.8
		7.5
	8.2	8.2
		9.1
10.0	10.0	10.0

The nominal ratings for passive electrical components (resistors, capacitors, inductors, Zener diodes) are described with the so-called *E-series* as per IEC 60063 (Preferred series for the nominal ratings of resistors and capacitors). The range covered is from $E3$ to $E192$ ($E3$: $r = 3$, i.e., three values per decade, $E6$: $r = 6$, $E12$: $r = 12$, etc.).

The most common series are $E6$, $E12$, and $E24$, subdividing the interval from 1 to 10 (or any other decade) into 6, 12, and 24 steps. Components from the lower E-series should be chosen for electronic designs, as they support lower inventory carrying costs.

Use of the E-series is mostly restricted to resistors, capacitors, inductors, and Zener diodes. Dimensions for other types of electrical components, such as fuses, are either chosen from the Renard series instead or are defined in relevant product standards.

8.4   Schematic Symbols of Electronic Components

Symbols in circuit diagrams are designated with an *identification letter* depending on the component type, followed by a consecutive number, (e.g., C4—capacitor no. 4, D1—digital gate no. 1, R12—resistor no. 12). The *value* of the component is also quoted with generic components, such as resistors, capacitors, and coils (e.g., 2.2 μ [F]), whereby the units are left out. One writes 2.2 μ instead of 2.2 μF for a capacitor, for example. The *type* should be specified, as it appears in the components library (e.g., NAND gate 74ACT00) for other electronic components, such as transistors or gates.

There are two sets of symbols for elementary logic gates in common use, one defined in ANSI/IEEE Std 91-1984[4] (supplement ANSI/IEEE Std 91a-1991) and IEC 60617,[5] respectively, and a "distinctive shape" set, based on traditional schematics. The latter is used for simple drawings and derives from military standards of the 1950s and 1960s.

Example: The symbol for a NAND gate in IEEE/ANSI/IEC standard (left) and its distinctive shape (on the right):

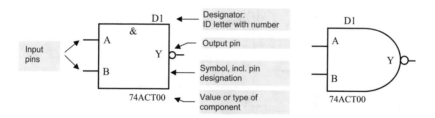

There are often a number of gates in a circuit enclosure. They are represented by a number of symbols with the same identification letter and the same number that refer to a common circuit enclosure. The example below shows four 2-input NAND gates in an enclosure D1. The individual gates are designated with D1A, D1B, D1C, and D1D (also: D1.A, D1.B, etc.) in the circuit diagram:

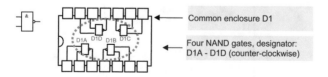

---

[4]IEEE Standard 91-1984, IEEE Standard Graphic Symbols for Logic Functions, and IEEE Standard 91a-1991, Supplement to IEEE Standard 91-1984.

[5]IEC 60617 Graphical symbols for diagrams.

The pins on a logic IC, for example, are typically numbered counter-clockwise, starting at the mark (pin assignment as per spec sheet, + supply voltage, ⊥ ground connection, IC as seen from above, mark is visible):

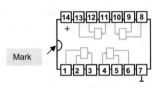

IEEE/IEC symbol	Description	Alternative/ distinctive shape
	Primary cell, secondary cell, storage battery The longer line represents the positive pole, the shorter one the negative pole.	
⊥ ▽ ⊥	Ground symbols	⊥ ▽ ⊥
	DC voltage source DC current source	
	Resistor	
	Resistor, adjustable	
	Capacitor Polarized capacitor	
	Inductor, coil, winding, choke (without core)	
	Semiconductor diode; (triangle = anode, bar = cathode)	
	Light emitting diode (LED) Photo- diode	

IEEE/IEC symbol	Description	Alternative/ distinctive shape
	pnp- npn- transistor	
	n-channel p-channel Junction FET	
	n-channel p-channel Enhancement MOSFET	
	n-channel p-channel (n-MOS) (p-MOS) Enhancement MOSFET (digital representation)	
	n-channel p-channel Depletion MOSFET	
	Photo transistor (npn model)	

Note: The transistor designations B, E, C and G, S, D mark the base, emitter and collector as well as gate, source and drain, respectively; they are not part of the actual symbol.

IEEE/IEC symbol	Description	Alternative/ distinctive shape
1	Inverter	
&	AND gate	
&	NAND (AND with negated output)	
≥1	OR gate	
≥1	NOR (OR with negated output)	
S D Q C Q̄ R	D-Flipflop	S D Q C Q̄ R
▷ ∞ - +	Operational amplifier	- +

8.5   Labeling of Electronic Components

8.5.1   Labeling with Colors

The values or ratings of resistors and inductive components are often indicated with color bands. The band nearest the edge is read first; there might be also a gap before the tolerance band to indicate the reading sequence. The first two or three color bands encode the digits of the nominal value, the next band the multiplier, and the last one the tolerance. The value is defined in ohms for resistors, and in microhenries for inductors.

International standards are IEC 60062 (Marking codes for resistors and capacitors) and IEC 61605 (Fixed inductors for use in electronic and telecommunication equipment—Marking codes).

Color code	Digits of nominal value (two or three bands)	Multiplier (one band)		Tolerance (one band)	
		Resistance	Inductance	Resistance (%)	Inductance
Silver	–	10^{-2}		± 10	
Gold	–	10^{-1}		± 5	
Black	0	1		–	$\pm 20\%$
Brown	1	10^1		± 1	
Red	2	10^2		± 2	
Orange	3	10^3		± 0.05	–
Yellow	4	10^4		± 0.02	–
Green	5	10^5		± 0.5	–
Blue	6	10^6		± 0.25	–
Violet	7	10^7	10^{-3}	± 0.1	–
Gray	8	10^8	10^{-4}	± 0.01	–
White	9	10^9	–	–	–
None	–	–		± 20	–

Examples:

A resistor with bands of brown, green, red, and silver: $15 \cdot 10^2 \ \Omega, \pm 10\% = 1.5 \ k\Omega, \pm 10\%$

A resistor with bands of violet, green, black, red, and brown:
$750 \cdot 10^2 \ \Omega, \pm 1\% = 75 \ k\Omega, \pm 1\%$

8.5.2 Labeling with Characters

8.5.2.1 Labeling Resistors and Capacitors with Digits and Letters

Resistors and capacitors can also be labeled with a combination of two, three or four digits for encoding the nominal value and a letter for encoding the multiplier. The position of the multiplier defines the position of the decimal point (IEC 60062, Marking codes for resistors and capacitors).

Resistor		Capacitor	
ID letter	Multiplier	ID letter	Multiplier
R	1	p	10^{-12}
K	10^3	n	10^{-9}
M	10^6	μ or u or U	10^{-6}
G	10^9	m	10^{-3}

(continued)

(continued)

Resistor		Capacitor	
T	10^{12}	F	1
Examples: 2M2 – 2.2 MΩ 68K0 – 68.0 kΩ 10R00 – 10.00 Ω		1F0 – 1.0 F 220µ – 220 µF 10n00 – 10.00 nF	

8.5.2.2   Labeling Resistors and Capacitors with Digits

Resistors and capacitors can also be labeled with three or four digits. The last digit encodes the multiplier. The value for resistors is specified in ohms, and in pico-farads for capacitors (IEC 60062, Marking codes for resistors and capacitors). Microfarads are typically used for capacitors with large capacitance values (e.g., electrolytic capacitors).

Resistor	Capacitor
Two or three digit nominal value, followed by a one digit multiplier	Two digit nominal value, followed by a one digit multiplier
Examples: 1200 – 120 · 10^0 Ω = 120 Ω 473 – 47 · 10^3 Ω = 47 kΩ	100 – 10 · 10^0 pF = 10 pF 223 – 22 · 10^3 pF = 22 nF

8.5.2.3   Labeling Inductances with Digits and Letters

Inductances with a value less than 10 µH are labeled with a combination of two digits and one letter. The letter encodes the multiplier and defines the position of the decimal point. From 10 µH upwards, the value is indicated by three digits. The last digit encodes the multiplier. The value is always specified in microhenries (IEC 61605, Fixed inductors for use in electronic and telecommunication equipment— Marking codes).

Value range	ID letter	Multiplier
<100 nH	N	10^{-3}
100 nH $\leq x <$ 10 µH	R	1
$\geq$ 10 µH	Two digit nominal value followed by a one digit multiplier	
Examples	4N7 – 4.7 · 10^{-3} µH = 4.7 nH R33 – 0.33 µH = 330 nH 10R – 10 µH 471 – 47 · 10^1 µH = 470 µH	

Index

0-9
0 V potential, 157
19-inch rack system, 36

A
Absolute zero, 80
Aging, 57
Air resistance (flow channel), 141
Appliance classes, 40
Availability, 65
  achieved, 65
  inherent, 65
  operational, 65

B
Back-annotation data, 24
Baffle, 136
Bar chart, 17
Bare die, 38
Base failure rate, 57
Bathtub curve, 52
Bill of components, 18
Black body (radiation), 101
Box assembly, 207
Box assembly technique, 37
Built-in obsolescence, 199
Bypass effect, 168

C
CAD model, 24
Capacitive coupling, 152
Cascade principle (recycling), 212
Celsius, 80
CE marking, 40
Chassis, 160
Chassis system, 37
Chimney effect, 136
Circuit diagram, 22

Circuit schematic, 22
Circular economy, 196
Coaxial cable, 176
Coextrusion, 212
Communication function, 34
Communication layer (of electronic systems), 33
Compact design, 35
Compatibility matrix, 213
Component (of electronic systems), 31, 35
Computer-aided design, 24
Conceptual stage, 7
Conduction heat transfer, 90
Conduction thermal resistance, 93
Conductive coupling, 150
Convection, 94
  forced, 94
  natural, 94
Convection heat transfer coefficient, 96
Convection thermal resistance, 96
Convective heat transfer, 94
Coupling, 148
  capacitive, 152
  conductive, galvanic, 150
  electromagnetic, radiative, 156
  inductive, magnetic, 154
Coupling coefficient, 155
Critical path (network plan), 17
Culprit (EMC), 148
Custom-assembled design, 35
Cutlery tray technique, 207

D
Datum (technical drawing), 226
Decoupling capacitor, 152
Derating, 60
Design architectures (of electronic systems), 35
Design for adaptability, 203

© Springer International Publishing AG 2017
J. Lienig and H. Bruemmer, *Fundamentals of Electronic Systems Design*,
DOI 10.1007/978-3-319-55840-0

Design for disassembly, 208
Design for durability, 202
Design for regeneration, 202
Design for reliability, 64
Design process, 6
Design stages (of a design process), 7
Desktop standing device, 38
Development stage (product), 5
Differential signal transmission, 165
Dimensioning (technical drawing), 21
Dimension line (technical drawing), 224
Dimensions (technical drawing), 224
Discrete component, 38
Disturbance layer (of electronic systems), 33
Downcycling, 212
Drift, 57
Drift fails, 71
Durability (electronic system), 201
Dynamic field, 168

E
Early failure, 52
Earth current, 164
Electrical energy, 79
Electromagnetic compatibility (EMC), 148
Electromagnetic coupling, 156
Electromagnetic field, 156, 168, 176
Electronic functional groups, 38
Electronic system, 31
Electronic systems design, 1
Electroquasistatic field, 168
Electrostatic discharge (ESD), 181
Electrostatic discharge protected area (EPA),
    183
Electrostatic field, 168
Elements (of electronic systems), 34
EMC-compliant system cabinet, 188
Emissivity, 102
Enclosure temperature, 117
Energy-storage capacitor, 152
Entropy (of materials), 194
Environment (of electronic systems), 32
ESD-suppression measures, 182
E-series (preferred numbers), 230
EU declaration of conformity, 40
Exponential distribution, 54
External electrical interconnects, 39

F
Failure density function, 51
Failure distribution function, 50, 55
Failure in time (FIT), 59
Failure rate, 51, 58
Fan, 110

axial-flow fan, 110
centrifugal fan, 110
tangential fan, 111
Fan acoustic noise, 112
Fan curve, 111, 142
Faraday cage, 174
Feasibility study, 12
Ferromagnetic materials, 169
FE simulation, 27
Finite-element model, 27
Finite-element simulation, 27
Floor standing device, 38
Form tolerance (technical drawing), 227
Functional specification, 13
Function-based dimensioning, 225
Function (of electronic systems), 32, 33

G
Gantt chart, 17
Gaussian normal distribution, 53
General tolerance (technical drawing), 226
Grashof number, 97
Gray body (radiation), 101
Ground, 157
    loop, 164
    multi-point, 160
    single-point, star shaped, 159
Ground bounce, 158

H
Hazard rate/function, 52
Heat, 80
Heat energy, 80
Heat exchanger, 138
Heat flow, 80
Heat flux, 80
Heat loss, 80
Heat pipe, 112
Heat sink, 107
Heat source (thermal network), 86
Heat transfer, 80

I
Identification number (technical drawing), 220
Immunity (EMC), 148
Implementation stage (of a design process), 7
Induction (ESD), 181
Inductive coupling, 154
Influence factors, 58
Ingress protection (IP) marking, 42
Integrated circuit (IC), 38
Interference (EMC), 148
International protection (IP) marking, 42
IP code, 42

ISO tolerance (technical drawing), 227

J
Junction temperature, 115

K
Kelvin, 80
Kirchhoff's law, 101

L
Label (circuit diagram), 22
Labeling of electronic components
    colors, 234
    letter, 235
Laminar flow, 94
Late failure, 53
Layered assembly technique, 37
Layout (of a circuit), 24
Leasing (electronic system), 201
Letter (circuit diagram), 22
Life cycle assessment, 215
Life cycle (electronic system), 198
Line style (technical drawing), 221
Line width (technical drawing), 221
Loading factors, 58

M
Magnetic coupling, 154
Magnetoquasistatic field, 168
Magnetostatic field, 167
Maintenance (circular economy), 199
Maintenance (reliability), 49
Marketing stage (product), 5
Material compatibility, 213
Material labeling, 215
Material recycling, 199
Mean time between failures (MTBF), 50, 56
Mean time to failure (MTTF), 50, 56
Minimum life-time, 65
Modular design, 36
Module (of electronic systems), 31, 35
Multi-chip module (MCM), 38
Multi-point ground, 160

N
Nested assembly technique, 37
Network nodes (thermal network), 86
Network plan, 15
Neutral conductor (N), 160, 189
Nominal dimension (technical drawing), 225
Normal distribution, 53
Nusselt number, 96

O
Operating point (fan), 111
Operating temperature of components, 115
Orthographic projection, 18
Overtemperature, 82

P
Panel-mounted device, 38
Paper size (technical drawing), 221
Parallel system/structure (reliability), 63, 68
Parasitic oscillation, 164
Partition panel effect (radiation), 105
Peltier effect, 113
Peltier element, 114
Perforation coefficient, 121
Permeability, 169
Pin assignment (IC), 232
Planned obsolescence, 199
Portable device, 38
Positional tolerance (technical drawing), 227
Power dissipation, 80
Power-supply elements, 39
Prandtl number, 97
Preferred numbers, 228
Printed circuit board (PCB), 39
Probability, 48
    random event, 47
    relative frequency, 47
Probability density function (PDF), 51
Probability of occurrence, 48
Processing function, 33
Processing layer (of electronic systems), 32
Production-based dimensioning, 225
Production waste recycling, 198
Product life cycle, 5
Product planning, 12
Product recycling, 199
Product requirement document, 13
Project structure plan, 15
Protection classes, 40

Q
Quasi-static field, 168

R
Radiation heat transfer, 98
Radiation heat transfer coefficient, 104
Radiation thermal resistance, 104
Random event (probability), 47
Random failure, 52
Rate of occurrence of failures (ROCOF), 65
Receptor (EMC), 148

Recycling, 193
    material, 199
    product, 199
Recycling code, 215
Recycling loop, 198
Reduction factors, 58
Redundancy, 63, 68
    cold, 63
    hot, 63
Reference conductor, 189
Reference ground, 157
Reference stress, 57
Regenerability, 202
Relationships (of electronic systems), 34
Reliability, 45, 49
    cost, 46
Reliability function, 50, 55
Renard series, 229
Repair (reliability), 49
Reparability, 202
Return line/conductor, 162, 184
Reynolds number, 97
Right-angled parallel projection, 18
Ripcord technique, 207
Roughness (surface), 227

S
Sandwich assembly technique, 37
Scale (technical drawing), 19, 220
Scheduling, 15
Schelkunoff impedance concept, 178
Schematic (of a circuit), 22
Sectional view (technical drawing), 20, 222
Security function, 34
Seebeck effect, 113
Serial system/structure (reliability), 63, 66
Set of drawings (of a system), 18
Shielding, 165
    $\lambda/10$ rule, 180, 188
    absorption loss, 178
    electromagnetic field, 176
    electroquasistatic field, 175
    electrostatic field, 173
    magnetoquasistatic field, 170
    magnetostatic field, 168
    reflection loss (EMC), 178
Shielding effectiveness (SE), 166
Shielding factor (S), 166
Simulation
    dynamic, 27
    finite-element, 27

Single-point ground, 159
Skin depth, 172
Skin effect, 171
Source of disturbance (EMC), 148
Specific thermal conductance, 91
Stack assembly technique, 37
Standard module design, 36
Static field, 167
Stress factors, 58
Structural correctness, 206
Structure (of electronic systems), 32, 34
Suitability for disassembly (design), 204
Suitability for disposal (design), 213
Suitability for recovery (design), 211
Suitability for separation (design), 210
Suitability of materials, 204
Suitability of quantities, 210
Surface roughness, 227
Surface specification (technical drawing), 227
Surface temperature (enclosure), 117
Survival function, 50
Symbol (circuit diagram), 22
System cabinet, 189
System ground (G), 157, 189
System impedance curve, 111, 141
System levels, 38
System pressure curve, 111, 141

T
Task definition, 12
Technical drawing, 17
Technical requirements document, 14
Temperature, 80
Temperature limit, 82
Temperature source (thermal network), 87
Test-based dimensioning, 225
Thermal capacity, 81
Thermal conduction, 90
Thermal conductivity, 91
Thermal convection, 94
Thermal energy, 79
Thermal glue, 110
Thermal grease, 110
Thermal interface material, 110
Thermal network, 84, 86
Thermal radiation, 98
Thermal resistance, 81
Thermosyphoning, 113
Title block (technical drawing), 219
Tolerances (technical drawing), 22
Tolerance, tolerancing, 224

Transfer impedance, 176
Triboelectric effect, 181
Turbulent flow, 94

U
Unreliability function, 50
Upcycling, 212
Useful life (electronic system), 199

V
Victim (EMC), 148

View factor, 107
Views (technical drawing), 18
Volumetric flow rate (fan), 141

W
Waste management, 196
Wearout failure, 53
Weibull distribution, 54
White body (radiation), 101